A Volume in The Laboratory Animal Pocket Reference Series

The Laboratory
HAMSTER &
GERBIL

The Laboratory Animal Pocket Reference Series

Series Editor
Mark A. Suckow, D.V.M.
Freimann Life Science Center
University of Notre Dame
Notre Dame, IN

Published Titles

The Laboratory Rabbit

The Laboratory Guinea Pig

The Laboratory Rat

The Laboratory Hamster and Gerbil

The Laboratory Cat

The Laboratory Small Ruminant

The Laboratory Swine

The Laboratory Mouse

A Volume in The Laboratory Animal Pocket Reference Series

The Laboratory
HAMSTER &
GERBIL

Karl J. Field, D.V.M., M.S.
Diplomate ACLAM
Hoffmann-La Roche Inc.
Nutley, New Jersey

Amber L. Sibold
Novartis Pharmaceuticals, Inc.
East Hanover, New Jersey

Editor-in-Chief
Mark A. Suckow, D.V.M.

CRC Press
Taylor & Francis Group
Boca Raton   London   New York

CRC Press is an imprint of the
Taylor & Francis Group, an **informa** business

Published in 1999 by
CRC Press
Taylor & Francis Group
6000 Broken Sound Parkway NW, Suite 300
Boca Raton, FL 33487-2742

© 1999 by Taylor & Francis Group, LLC
CRC Press is an imprint of Taylor & Francis Group

No claim to original U.S. Government works
Printed in the United States of America on acid-free paper
10 9 8 7 6 5 4 3
International Standard Book Number-10: 0-8493-2566-8 (Softcover)
International Standard Book Number-13: 978-0-8493-2566-3 (Softcover)

Library of Congress Cataloging-in-Publication Data

Catalog record is available from the Library of Congress

**Visit the Taylor & Francis Web site at
http://www.taylorandfrancis.com**

**and the CRC Press Web site at
http://www.crcpress.com**

dedication

K.J.F.: As usual, I needed a lot of help completing this project. Thank you to all of the dedicated animal technicians who have devoted their lives to providing compassionate care for the animals they oversee. Thanks to my parents, William and Virginia, and to my mentors and friends in the profession. Lastly, I must thank my wife Nanette and daughters Lindsay and Whitney for their love and support.

A.L.S.: I would like to thank my parents, Faith and Jack, my son Randy and friend David, and all my talented co-workers for their support in this and all my endeavors.

preface

This handbook is intended to be a resource for a wide variety of institutions. However, individuals at small institutions or facilities lacking a large, departmentally based animal resource unit, and individuals who need to develop a research program for hamsters and gerbils from scratch, may gain the most benefit from this handbook. In smaller institutions, individuals who perform animal research often hold the responsibilities associated with animal facility management, animal husbandry, regulatory compliance, and performance of technical procedures directly related to research projects. It is our hope that these individuals in particular — the investigators, technicians and animal caretakers who are charged with the care and/or use of hamsters and gerbils in the research setting will benefit from the user-friendly format in which this handbook is written.

This handbook is organized into six chapters: Important Biological Features (Chapter 1), Husbandry (Chapter 2), Management (Chapter 3), Veterinary Care (Chapter 4), Experimental Methodology (Chapter 5), and Resources (Chapter 6). Basic information and common procedures which fall under the chapter scope are presented in detail. However, information such as alternative techniques, or details of procedures and methods which are beyond the scope of this handbook, are referenced. The references direct the user to additional information without having to wade through voluminous detail here. This handbook should be viewed as a basic reference source and not as an exhaustive review of the biology and use of the hamster and gerbil.

The final chapter on "Resources" provides the user with lists of possible sources and suppliers and additional information on hamster and gerbils, as well as sources for feed, sanitation supplies, cages, and research and veterinary supplies. The lists are not exhaustive and does not imply endorsement of listed suppliers over those not listed; they are a starting point for users to develop their own lists of preferred vendors of such items. Chapter 6 contains tables which list vendors of cages and research and veterinary supplies by number and is followed by a list of contact information for these suppliers.

A final critical point is that each individual involved in animal care and use must recognize that the humane care and use of hamsters and gerbils is a fundamental principal of good research. It is the principal investigator's obligation to ensure that each individual who is required to perform the procedures described in this handbook is properly trained, otherwise the animal may be harmed. An individual's inexperience with a technique can represent one of the most critical variables in a research project. The initial and continuing education of all personnel involved in the care and use of animals will facilitate the overall success of any program using hamsters and/or gerbils in research, teaching, or testing.

the authors

Karl J. Field, DVM., MS., is Director of Veterinary Sciences at Bristol-Meyers Squibb Pharmaceutical Research Institute in Princeton, NJ.

Dr. Field earned the degree of Doctor of Veterinary Medicine from Michigan State University in 1986 and completed an M.S. in laboratory animal medicine in 1988. He is a Diplomate of the American College of Laboratory Animal Medicine.

Dr. Field is the author of 20 publications and abstracts covering a variety of aspects of laboratory animal science.

Amber L. Sibold is Assistant Director of Laboratory Animal Services at Novartis Pharmaceuticals, East Hanover, NJ. Ms. Sibold recieved her undergraduate degree from Penn State University. She is a Certified Laboratory Animal Technologist and a Surgical Research Specialist.

contents

important biological features

The Syrian hamster and Mongolian gerbil have been used extensively in research for a variety of disciplines. Therefore, unless noted otherwise, the information in this handbook will focus on the Syrian hamster (Figure 1), the most commonly used hamster species in the laboratory,[1] and the Mongolian gerbil (Figure 2).

hamsters

Origin of the Hamster

The hamster has a short history of domestication. Most hamsters used as laboratory animals appear to be descendants of 3 to 4 littermates that were captured and imported from Aleppo, Syria in 1930. Because of their origin in Syria and their reddish golden-brown color, they are often referred to as the Syrian hamster, golden hamster, or golden Syrian hamster. The Syrian hamster belongs to the taxonomic order *Rodentia*, family *Cridetidae*. The genus and species is *Mesocricetus auratus*.[2]

Hamster Breeds

As mentioned previously, the Syrian hamster, *Mesocricetus auratus*, is the most frequently used hamster species in the

1

Fig. 1. An adult Syrian hamster, *Mesocricetus auratus*. (Courtesy of Mr. J. Foster, Jr., Charles River Laboratories.)

Fig. 2. An adult Mongolian gerbil, *Meriones unguiculatus*.

laboratory. Less commonly used breeds include the Striped or Chinese hamster, *Cricetulus griseus;* the European hamster, *Cricetus cricetus;* the Djungarian hamster, *Phodopus sungorus;* and the Armenian hamster, *Cricetulus migratorius.* A more complete listing of hamster breeds and classifications can be found elsewhere.[1,3]

Hamster Behavior

Although hamsters are not naturally aggressive toward their handlers,[1] they will display this behavior if they are startled, suddenly awakened, or roughly handled. Similarly, an unreceptive breeding female may bite when handled. Since hamsters are nocturnal and usually sleeping when first approached, it is important to awaken the hamster before you handle it.

Hamsters are compatible within their own species, regardless of sex, if they have been weaned and raised together. As adults, hamsters are aggressive towards unfamiliar animals of their own and the opposite sex when the new animal is introduced into their group, with the exception being the female in heat. A female in heat will usually be receptive to an unfamiliar male.[1]

Hamsters are very busy during their active period at night, exhibiting wheel-running activity and will climb on the cage top and through plastic tunnels (Figure 3). They are escape-prone so the cage lid must fit securely or they will push their way out. Hamsters will hoard food in their cheek pouches and at specific sites within their cages.[7]

Anatomic and Physiologic Features of Hamsters

- Female breeding hamsters are more dominant, aggressive, and larger in size than the male.

- The adult female is larger than the adult male.

- The vertebral formula is 7 cervical; 13 thoracic; 6 lumbar; 4 sacral; 13 to 14 caudal vertebrae.

- "Flank organs" are sebaceous glands that are located on the hip region in both males and females.[5] Flank organs are responsive to androgens and become coarse, darkly pigmented, and larger in males than in females. They are used for territorial marking.[6]

Fig. 3. A hamster inside a PVC pipe fitting. Objects in the cage that hamsters can climb in and through provide enrichment.

- Although not true hibernators, hamsters will begin to gather food and enter into pseudohibernation as the days shorten and the temperature decreases to 5°C to 15°C.[7]

- The dental formula is 2 (incisors 1/1, canines 0/0, premolars 0/0, molars 3/3). Incisors and molars are open rooted.

- The crowns of molars retain fine food, making hamsters prone to dental caries. Males are more prone to caries than females.

- The major salivary glands are the submaxillary, parotid and sublingual glands. A smaller retrolingual gland is associated with the submaxillary gland.

•   Buccal cheek pouches (Figure 4) are located along the lateral side of the head and neck region. They are used to hoard and carry food. They are occasionally used for tumor implant studies and are believed to be immunologically protected.[5] If startled, a female will hide her young in her cheek pouches. The hamster's stomach is compartmentalized into two distinct regions: the nonglandular forestomach and the glandular stomach (Figure 5) The forestomach is lined with keratinized epithelium, much like the forestomach of a ruminant, has a higher pH than the glandular stomach, and contains microflora and volatile fatty acids much like those found in the rumen of sheep and cattle. The esophagus empties into the forestomach. The glandular stomach is similar to that found in monogastric animals.[7]

•   The hamster has a single lobe on the left side of the lungs and 4 lobes on the right side.

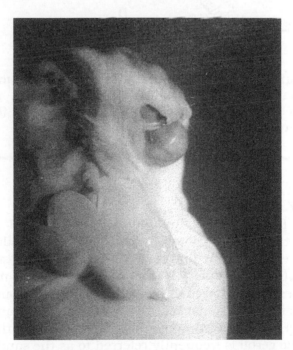

Fig. 4. An everted hamster cheek pouch. (Photo courtesy of Dr. Susan Gibson, University of Southern Alabama.)

- The hamster has small and fairly inaccessible superficial veins, and no tail vein, making blood sampling difficult. The hamster has an orbital venous sinus similar to the mouse.

- Hamsters, like mice and rats, have muscle fibers analogous to the myocardium that extend into and along the walls of the pulmonary vein.[5]

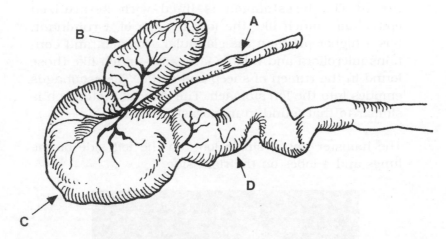

Fig. 5. Anatomy of the hamster stomach. The structures are the esophagus (A), nonglandular forestomach (B), glandular stomach (C), and the pyloric region of the stomach (D), which empties into the small intestine.

gerbils

Origin of the Gerbil

Gerbils (see Figure 2) are found in desert and semi-arid regions of the world, particularly northern Africa, India, southwest and central Asia, northern China, Mongolia, and some areas of Eastern Europe.[8] The Mongolian gerbil, the most common species of gerbil used in research, originated from 20 pairs of animals that were imported from eastern Mongolia to Japan, of which 11 pairs were eventually imported into the United States.[9] The gerbil is frequently referred to as the sand rat, jird, or desert rat. The name "jird" is often reserved for gerbils of the genus *Meriones*, and most frequently used in reference to the

Mongolian gerbil. The Mongolian gerbil, like the hamster, belongs to the order *Rodentia*, family *Cridetidae*. The Latin genus species name for the Mongolian gerbil is *Meriones unguiculatus*.[10]

Gerbil Breeds

About 100 breeds of gerbils have been described,[11] ranging in size from that of a mouse to that of an adult Norway rat. As mentioned previously, the most common gerbil used for research is the Mongolian gerbil, *Meriones unguiculatus*. Other breeds mentioned in gerbil fancier magazines, and occasionally the scientific literature, include *Meriones shawi*, *Meriones libycus*, *Meriones persicus*, *Gerbillus amoenus*, and *Gerbillus pyramidum*.

Gerbil Behavior

Gerbils are gentile and docile by nature,[8] intensely curious, will use plastic enrichment devices to climb on and play in (Figure 6), and have a reputation for sticking their nose into

Fig. 6. Gerbils are intensely curious and, like hamsters, will express their exploratory behavior and gain enrichment through the addition of enrichment devices to their cage.

everything. They live in colonies in their natural environment, naturally burrow, and are known to constantly scratch the cage floor with their front and hind paws. Gerbils have been known to damage the floor of a plastic cage from their burrowing behavior. Most gerbils are active throughout the 24-hour period, with only slight differences during the dark phase.[10] In the wild, there is little activity during the cold nights of spring and fall. Both sexes will naturally hoard food, with the female more prone to hoarding than the male.

Gerbils will form a strong pair bond, mating for life and breeding throughout the year.[11] A harem mating setup of one male to two females is often used by breeders in production colonies.[12] This pairing should be established after gerbil pups are weaned but before puberty, which occurs at approximately 6 to 7 weeks of age.[13] Gerbils that have been separated and reunited after reaching sexual maturity will often fight. Introduction of a new mating partner after one of the original partners dies often leads to fighting and death of the new member.

Gerbils will greet one another by licking each others' mouth, are excellent jumpers, and thump with their hind limbs when frightened. When they are ready to fight, they will push each other with their heads and begin to box and wrestle.

Anatomic and Physiologic Features of Gerbils

- Gerbils will burrow through their bedding with their noses. This may predispose them to a syndrome known as "sore nose" (see Chapter 4).

- If gerbils are handled by the tip to middle of the tail, they may slough their tail.

- Both sexes of gerbils will build a nest and care for the young.

- The dental formula is 2 (incisors 1/1, canines 0/0, pre-molars 0/0, molars 3/3).

- Both sexes of gerbils are prone to epileptiform seizures. There are seizure-prone and seizure-resistant strains available.

- Both male and female gerbils have a mid-ventral abdominal sebaceous gland. The male sebaceous gland is twice the size of the female's gland, and may be mistaken for a tumor. The male rubs this gland against objects to mark his territory, a behavior that is androgen dependent. Gerbils with a black coat color tend to mark territory more frequently than those with a brown coat.[14]

- After parturition the female will mark her territory and become aggressive.

- The gerbil's adrenal-to-body-weight ration is 3 to 4 times that of the rat.[15,16]

- Approximately 40% of gerbils have an incomplete Circle of Willis. As a result, ligation of one carotid artery will result in a cerebral infarct on the side ipsilateral to the ligation.[17]

- Gerbils are prone to dental caries and periodontal disease. The incidence of caries increases after 6 months of age.

- Gerbils are used extensively in research involving aging, audiology, dental disease, endocrinology, oncology, nutrition, radiobiology, reproduction, infectious diseases, and strokes.

normative values

Typical values are presented for basic biologic parameters (Table 1), clinical chemistry (Table 2), cardiovascular and respiratory function (Table 3), and hematology (Table 4). The values are representative of those in the Syrian hamster and Mongolian gerbil. The normative values found in the tables can vary significantly due to factors such as differences between individual animals, breeds, age, sex, housing conditions, disease status, laboratories, and sampling methods, therefore, these tables are provided as guidelines. It is imperative that each laboratory establish their own normal reference values. Values are presented as an average or a range from data given in References 1, 5, 13, 18, 19, 20, 21, 22, 23, 24, and 25.

TABLE 1. BASIC BIOLOGICAL PARAMETERS OF THE HAMSTER AND GERBIL

Parameter	Hamster	Gerbil
Chromosome number	44	44
Life span (years)	1-3	2-5
No. mammary glands	6-11 pair	2 pairs: 1 thoracic, 1 inguina
Birth weight (g)	1-2	3-4
Mature body weight (g)	85-130 (M) 95-150 (F)	65-130 (M) 55-133 (F)
Body temperature (°C)	37-38.5	37-39
Dental formula	2(1/1, 0/0, 0/0, 3/3)	2 (1/1, 0/0, 0/0, 3/3)
Food intake (g/day)	10-14	5-7
Water intake (ml/day)	7-10	4-7
Urine volume (ml/day)	6	3-4
Diet peculiarities	Feed on cage floor	Prefer seeds to hard food

TABLE 2. CLINICAL CHEMISTRY VALUES OF THE HAMSTER AND GERBIL

Parameter	Hamster	Gerbil
Total protein (g/dl)	4.3–7.7	4.3–12.5
Globulin (g/dl)	2.08–6.28	1.2–6.0
Albumin (g/dl)	2.63–4.10	1.8–5.5
Acid phosphatase (IU/L)	3.9–10.4	NV
Alkaline phosphatase (IU/L)	3.9–10.4	NV
LDH (IU/L)	148-412	NV
Aspartate aminotransferase (IU/L)	37.6–178	NV
Creatine kinase (IU/L)	0.50–1.90	NV
ALT (SGPT) (IU/L)	11.6-35.9	NV
Blood urea nitrogen (mg/dl)	15–33	17–32
Creatinine (mg/dl)	0.4–1.0	0.64–1.12
Glucose (mg/dl)	65–173	50–135
Sodium (meq/L)	106–146	143–157
Chloride (meq/L)	86–112	105
Calcium (mg/dl)	7.4–12.0	3.6–6.0
Magnesium (mg/dl)	1.9–3.5	NV
Potassium (meq/L)	4.0–5.9	3.9–5.2
Phosphorus (mg/dl)	3.4–8.24	3.7–7.1
Total bilirubin (mg/dl)	0.2–.74	1.3–2.5
Uric acid (mg/dl)	1.8–5.0	1.1–2.8
Amylase (IU/L)	120–250	NV
Serum lipids (mg/dl)	223–466	NV
Phospholipids (mg/dl)	132–190	NV
Triglycerides (mg/dl)	72–227	NV
Cholesterol (mg/dl)	112–210	90–151

TABLE 3. CARDIOVASCULAR AND RESPIRATORY FUNCTION OF THE HAMSTER AND GERBIL

Parameter	Hamster	Gerbil
Respiratory rate (breaths/min)	74 ave., 33–127	70–120
Heart rate (beats/min)	332 ave., 286–400	360 ave. 260–600
Tidal volume (ml)	0.6–1.4	NV
paO2 (mmHg)	67–77	NV
paCO2 (torr)	38.5–43–5	NV
HCO3- (mmol/L)	27–33	NV
Arterial blood pH	7.45–7.51	NV
Venous blood pH	7.33–7.39	NV

TABLE 4. HEMATOLOGICAL VALUES OF THE HAMSTER AND GERBIL

Parameter	Hamster	Gerbil
Packed cell volume (%)	39 59	44–49
Red blood cells (10^6/ul)	7.2 ave., 4–10	7–10
White blood cells (10^3/ul)	6.5–10.6	7.3–15.4
Hemoglobin (g/dl)	13–19	13–16
Neutrophils (%)	1.9–3.1	1.3–5.2
Lymphocytes (%)	4.4–7.2	5.1–11.8
Eosinophils (%)	0.0–.10	0.07–0.32
Basophils (%)	0.0–1.2	0.1–0.28
Monocytes (%)	0.4–4.4	0.03–0.25
Reticulocytes (%)	1.3–3.7	2.0-5.4
MCV (fl)	64–76	NV
MCHC (%)	28-37	NV
MCH (pg)	20–26	NV
Blood volume (ml/kg)	65–80	60–85

NV = No value found

notes

2

husbandry

The well-being of all laboratory animals is inseparably linked to the quality of management of the physical environment in the animal facility. Critical areas of concern include the animal's macro- and microenvironment, nutrition, sanitation, and minimization of factors which predispose the animal to disease and injury.

housing

The term macroenvironment refers to the environment within the room surrounding the microenvironment. The term microenvironment refers to the internal environment of the cage the animal is housed in. The physical environment within the microenvironment includes the intracage temperature, humidity, air exchange rate, ammonia and carbon dioxide concentrations, and illumination level. The components of the microenvironment will differ with changes in the type of cage provided to the animal (e.g., solid bottom with an open top and wire lid versus the same housing with a microisolator lid). For example, changes in the ventilation rate within the room can affect the humidity within the animal's cage.[26] The ability for air to be exchanged through the open lid will be greater than that when a microisolator lid is used.

Macroenvironment Considerations

room construction features

Hamsters and gerbils can be housed in animal rooms designed for rats and mice. As with any animal room, those housing hamsters and gerbils should be located in an area that has minimal noise. The room should have a close proximity to support areas associated with the research project that the animal is used in. This minimizes the need to transport the animals long distances to perform procedures.

- **Doors** to the animal room should be designed to prevent both the entry of vermin into the room and the exit of escaped animals out of the room. A self-closing door with a door sweep will prevent an escaped animal from leaving the room. *The Guide for the Care and Use Of Laboratory Animals* (the "Guide")[27] recommends that doors be 42 in. wide by 84 in. high, thus allowing easy movement of equipment into the room. The door should also be finished with a material that prevents and resists corrosion. Door jamb guards and 3/4 in. high kick plates are very helpful in preventing damage to doors. Cage bumper guards also assist in the protection of doors.

- **Exterior windows** are often considered undesirable because they may introduce additional light, which makes it difficult to control the temperature and the photoperiod in the room.

- **Floors and walls** should sealed with a water-repellant material that leaves the surface pinhole-free, minimizing areas to harbor bacteria. Walls are often painted with epoxy paint to achieve this effect. The floor and wall surfaces should also be nonabsorbent, impact-resistant, and resistant to deterioration from detergents, disinfectants, urine, and other materials that may contact them. Generally, floors should be smooth, allowing equipment to roll freely, and they should support the equipment and racks rolled across them without cracking. Commonly used flooring materials include urethanes; epoxies; sheet vinyl; acrylics such as methylmethacrylate; quarry tile; and sealed, hardened concrete. In a wet environment, the floor should be textured to avoid slippage, and should slope towards the drain. The slope should not cause racks to tip when being moved into the room, or to rest in an unleveled position after being placed in the room. Wall materials should be impact-resistant, and may include a bumper guard or rail, corner guards, and base cants to prevent damage from equipment. Commonly used materials include concrete block, gypsum wallboard, tile, or cement board. The wall surface is often coated with epoxy paint. All cracks, crevices, and joints should be sealed on all floors and walls.

- **Junctions** where the floor meets the wall and wall meets the ceiling should be sealed to prevent accumulation of dirt and debris. Consideration should be given to using a sealant or caulk that prevents the growth of bacteria and fungi.

- **Ceilings** should be smooth and coated with a material that withstands deterioration from detergents, disinfectants and other materials that may contact the surface. Typical ceiling finishes include gypsum wall board, fiberglass reinforced panels, and mylar-coated panels or a ceiling grid system comprised of fiberglass, baked enamel, stainless steel, or aluminum panels.

- **Environmental control** parameters will be covered in more detail later in this chapter. The room environment should be closely controlled and monitored so as to prevent fluctuations in the temperature, humidity, the light cycle, ammonia concentrations, and the hourly rate of air exchanges.

Microenvironment/ Caging Considerations

The microenvironment (primary enclosure) should conform to the following requirements: [27,28] it should be secure, structurally sound, free of sharp edges, and in good repair, thereby protecting the animal from injury and accidents. Wire-mesh floors, although not encouraged by the Guide, when used must allow feces to pass through, but protect the animals' feet and appendages from injury; animals must be able to maintain normal body temperature, remain clean and dry and have access to food and

Fig. 7. Assortment of multiple caging used for hamsters and gerbils. Both solid-bottom and wire-bottom cages are shown

water; the enclosure must allow for easy observation of the animals; and, the enclosure must allow animals to make normal movements and postural adjustments (Figures 7 to 10).

cage size standards

Hamster cage size specifications are included in both the Regulations of the Animal Welfare Act,[28] and the Guide.[27] Since the caging specifications differ between these two publications, it is important to recognize that the caging standards listed in the Animal Welfare Act are regulations set by the Federal Government (Tables 5 and 6).

Gerbil caging standards are not provided in either of the above references. Gerbils are typically housed in plastic or metal cages (see Figures 7 to 10) used for hamsters, mice, or rats. Solid flooring with bedding is preferred by gerbils. A plastic aquarium may also be used. Mature gerbils 12 weeks of age and older should receive 36 square inches of floor space per animal, while 180 square inches of floor space is sufficient for a breeding

Fig. 8. When hamsters are placed in wire-bottom cages, the spacing of the wires must be narrow enough so that their feet do not pass through and become injured.

Fig. 9. Typical shelf rack containing solid-bottom cages and an automatic watering system.

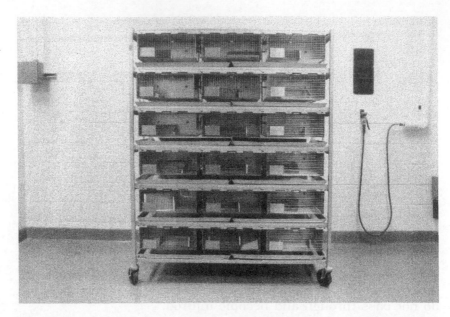

Fig. 10. Typical shelf rack containing wire-bottom cages and an automatic watering system. The unit shown is mechanically ventilated with self-contained intake and exhaust units.

pair.[29] A smaller cage may inhibit breeding. Since gerbils often sit up on their hind feet, the cage height provided to rats of 7 inches of internal height is adequate for an adult breeding pair.[15] Because gerbils are burrowing animals, wood chips, corncobs, cellulose-based bedding, sand, or wood pulp provides good bedding. Pine shavings or saw dust are not recommended because the fur will become greasy and will mat.

cage materials and design

Caging used to house other laboratory rodents are not acceptable for hamsters and gerbils unless they are escape-proof, and constructed of a material (e.g., polycarbonate, stainless steel) that prevents the hamster from gnawing through (Figure 11). Likewise, overhead sipper tubes used for rats and mice are often too high for weanling and suckling hamsters to reach.[7] Extended-length sipper tubes or automatic watering systems should be used in these instances.

Caging materials should be designed to be smooth and impervious to moisture and liquids, corrosion resistant, withstand deterioration during decontamination and provide adequate floor support for the animals housed within the enclosure. Although

TABLE 5.   MINIMUM CAGING REQUIREMENTS FOR HAMSTERS[a]

Age in weeks	bMinimum Space per Hamster, in²		Max. Population per Enclosure	Interior Height, in²
	Dwarf	Other		
up to 5	5	10	20	c5–5 1/2
5–10	7.5	12.5	16	"
>10	9	15	13	"

a Regulations of the Animal Welfare Act[28] for primary enclosures acquired before August 15, 1990.

b A nursing female with her litter must be housed in a primary enclosure which provides at least 121 square inches of floor space, provided, however, that in the case of Dwarf hamsters the floor space is at least 25 square inches.

c The cage interior height is a minimum of 5 1/2 inches, except for Dwarf hamsters, in which case it must be a minimum height of 5 inches.

TABLE 6.   CAGING SIZE GUIDELINES FOR HAMSTERS[a]

Body Weight, g	Floor Area Required per Hamster, in² (cm²)	Interior Cage Height, in²
<60	10 (64.5)	6
60–80	13 (83.9)	6
81–100	16 (103.2)	6
>100	19 (122.6)	6

a Summarized from the *Guide for the Care and Use of Laboratory Animals*[27] and the Animal Welfare Act.[28] These space requirements apply to primary enclosures acquired after Agust 15, 1990.

stainless-steel wire-bottom cages are often used, hamsters and gerbils prefer solid-bottom cages with contact bedding.[7,29,30] As mentioned before, hamsters have the ability to flatten their bodies and squeeze out of small spaces, and gerbils like to burrow and push with their face. Therefore, the cage lid of solid-bottom cages must secure firmly to the cage side to prevent escape. For wire-bottom cages, the juncture of the walls to the top of the cage must fit snugly to prevent animals from escaping.

Fig. 11. Escape-proof caging is required for hamsters and gerbils. Note that the lid locks down on this cage. This cage also contains a cage card, water bottle, and enrichment tubing

Solid-bottom caging with contact bedding, although more labor intensive to maintain, is a more natural environment for hamsters and gerbils. They instinctively burrow, grow more rapidly, and experience fewer stress-related deaths than animals housed in wire-bottom cages. The addition of tissues or cotton bedding material will also improve litter yields for hamsters.[31] If laboratory gerbils are exposed to a naturalistic environment containing items such as stones, sand, and a water dish, they will find food, rear up, mark their cage, and move about more freely.[29] Gerbils are known to damage cage flooring by constantly scratching the surface with their front paws.

Stainless-steel cages, when provided with a solid back and sides and a wire-mesh front, will allow hamsters to seek their own level of privacy in the darker corners of the cage. Hamsters on wire floors exhibit greater hoarding behavior and spend less time gnawing than solid-bottom housed animals.[30]

As mentioned before, the wire-mesh floor should have openings large enough to allow feces to pass through, while preventing the animals' feet and appendages from falling through and becoming injured. The gauge of wire used and width of mesh is often the same as that used for rats. It is important to note that hamsters have a tendency to sleep in piles. If wire-bottom cages are used, the gauge of the mesh should be sufficient to support the animal's weight without collapsing. Suspended cages are not acceptable in a breeding colony, or for mothers with a litter because the pups may fall through, and they may also develop hypothermia and die.[31]

Hamsters and gerbils adapt easily to automatic watering as well as water bottles. Since hamsters go to great extremes to remove food from overhead feeders, feed may be put on the floor.[28]

Environmental Conditions

The animal housing environment should be controlled to minimize variations in the temperature, humidity, and lighting levels. Otherwise, variations in these parameters may introduce unwanted variables into an experiment. Performance of the heating, cooling, ventilation, humidity, and lighting systems can be monitored through the use of a hand-held meter (Figure 12), physical room check for light activation and deactivation, or continuously for all three parameters through a centralized monitoring system.

The laboratory hamster is derived from a desert environment and the gerbil from a semi-arid environment, thus both adapt fairly well to the laboratory environment. In general, it can be said that the environmental conditions acceptable for rats and mice can be applied to hamsters and gerbils. Since hamsters are natural hibernators, entering hibernation at approximately 5°C, they adapt more easily to cold temperatures than warm temperatures.[31] Adult hamsters and gerbils are usually maintained at 18 to 29°C, and young at 22 to 24°C.[27,29] Hamsters will enter a daylight sleep and can be difficult to arouse at higher temperatures,[31] and will become heat stressed at 34°C and begin dying at 36°C.[29] Since hamsters are more active during the dark cycle, caution should always be used when arousing a hamster sleeping during the day. They have a tendency to be aggressive when awakened. In contrast, gerbils housed in a laboratory environment usually show activity throughout the 24-hour period, with only slight differences during the dark phase of the light cycle.[10]

For both hamsters and gerbils, relative humidity should range between 30 to 70%.[27] The light cycle is most frequently set at

Fig. 12. A hand-held temperature and humidity monitoring device can be used to measure the environmental parameters within the animal facility. This piece of equipment, as with other monitoring devices used within the animal facility, should be calibrated within a scheduled interval.

12:12 lights on:lights off cycle. For hamsters, a 14:10 on:off cycle is optimal for breeding colonies.[29] An excessively longer or shorter light cycle than what is recommended above will lead to infertility.[29]

Noise should be kept at a minimum within the animal room. Female hamsters with litters are particularly irritable, and sudden noises or disturbances will promote this behavior.[31] Both male and female gerbils may experience epileptiform seizures, characterized by convulsions and prostration lasting several minutes, in response to sudden noise.[10]

environmental enrichment

Hamsters are most active at night, although they will exhibit brief periods of activity during the daytime. During the daylight hours, they will seek the darkest area of the cage and sleep in

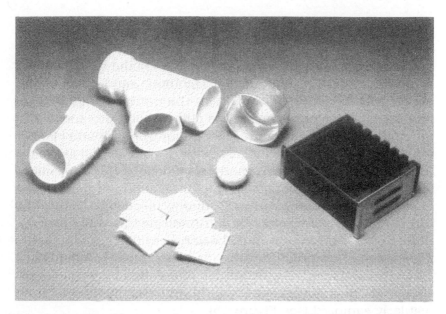

Fig. 13. Enrichment devices such as PCV tubing, feeders, balls, and cotton nesting material may be added to the cage.

piles. Hamsters enjoy exercise wheels and others toys such as small boxes, plastic or cardboard tubing, tin cans, and other objects which they can climb into and on (Figure 13). Whatever type of enrichment device is added to their environment, it should be non-toxic and sanitizable.

Gerbils are usually housed in pairs after sexual maturity. If they have been separated from one another for a period of time, or have been individually housed, they will fight when a pair bond or group housing is set up. This may be avoided by placing the gerbils in a neutral cage at the time the cage is changed, providing areas for newly introduced animals to hide (a cup, metal dish, etc.), or by forming the pair bond at approximately 10 weeks of age.[15] Gerbils, being natural burrowers, benefit from an environment that includes stones, sand, and other forms of bedding, as well as enrichment devices as described above for hamsters. Gerbils are very aggressive chewers and will destroy enrichment devices quickly.

nutrition

Hamsters. Although the hamster diet should be capable of supporting growth, reproduction, lactation, and adult maintenance, there is little information on the nutrient requirements for the hamster when compared to other laboratory rodents.[1] In the wild, hamsters are omnivorous, existing on fruit and plants.[7] Within the laboratory setting, hamsters are usually fed a pelleted rodent diet intended for rats and mice (Figure 14). Supplementing the diet with fruits and vegetables should not be necessary if the laboratory diet is of high quality. In addition, providing fruits and vegetables may be contraindicated as this may be a route of exposure to unwanted bacteria or other contaminants. Food should be stored properly at temperatures recommended by the manufacturer, with strict adherence to recommended shelf life. The storage area should be clean, uncluttered, and regularly sanitized (see Figure 15).

Fig. 14. Both natural and manufactured food products are commonly given to hamsters and gerbils. Natural food products often provide a source of both nutrition and water.

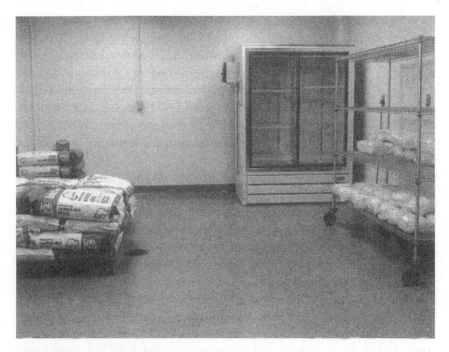

Fig. 15. Food storage should occur in a dedicated feed storage area. The food should be stored off the floor and the area should be sanitized regularly. The temperature and humidity should be set at a level that will maximize the shelf life and reduce spoilage of food, especially fresh fruits and vegetables.

The presence of the forestomach in the hamster is considered by some to be similar in function to that of a cow's rumen. This anatomical feature has created discussion on proper hamster nutrition,[32] with some authors speculating that a lower quality diet may be provided due to the hamsters' increased capacity to digest roughage. Nutrient levels for the hamster diet are usually found to be within the following range of formulation: crude protein of 17 to 23%; crude fat of approximately 4.5%; and crude fiber of 6 to 8.0%.[1,29] Of critical importance in the hamster diet is vitamin E. Failure to provide adequate vitamin E can result in deaths of fetuses and muscular dystrophy in weanlings.[29,31]

Hamsters begin to gnaw on pelleted food around 7 to 10 days of age, and as adults will eat around 7 to 15 grams of food per day and drink up to 10 ml of water/100 g body weight per day.[29] Food consumption is at the higher end of the scale in pregnant and lactating females. Lactating females also require increased water intake, otherwise milk production will decrease and the young will starve.[7] Hamsters are hoarding animals and will remove all the feed from a feeder and deposit it around the cage and in their cheek pouches. Due to this behavior, it is acceptable to place pelleted food on the floor of the hamster cage.[28]

Gerbils. Although little is known about the nutritional requirements of gerbils, they are known to be grainivorous or herbivorous. The recommended diet is a pelleted rodent chow providing 16 to 20% protein, fed *ad libitum* (gerbils eat approximately 8 times per day) in a feed bowl, dish, or hopper,[15,29,33] Their diet is often supplemented/enriched with a mixture of seeds, vegetables, and grains (Figure 16), although such foodstuffs are not nutritionally complete and should only be provided as a treat. Young gerbils begin eating at around 2 weeks of age, and may have difficulty getting to food if the food bowl is too high; it may be necessary to provide some food on the cage floor. Adult gerbils will consume 5 to 8 grams of food per day.[15,29]

Fig 16. Natural food provides a source of nutrition, water, and enrichment.

Gerbils are not coprophagic when fed a commercially available rodent diet *ad libitum.* However, they may become coprophagic when fed a nutritionally incomplete diet.[33]

Although in the natural environment gerbils obtain most of their water from their diet, in the laboratory the pelleted chow provides very little moisture. Therefore, adult gerbils housed in a laboratory environment must be provided with a continuous supply of clean water (see Figure 17), and will drink 4 to 7 ml water daily.

Fig. 17. When using an automatic watering system, the water line must be flushed when the rack is initially hooked up. This flushes all the air out of the system, ensuring that fresh water is distributed throughout the system.

cage sanitation practices

The goal of cage sanitation is to maintain the animals' living conditions at the highest level possible in order to promote optimal health and well being. There are three critical components of sanitation for the animal's primary enclosure: changes in bedding and water bottles, cage changes, and cage disinfection.

Cage Cleaning Frequency

For animals housed in solid-bottom cages, bedding changes are needed to separate the animal from exposure to feces and urine that build up in the cage. In addition, fresh bedding provides new, clean material to keep the animals clean and dry. Types of bedding used in solid-bottom cages include purified wood pulp, chopped corncob, hardwood, cellulose fiber (hamsters and gerbils), and sand (gerbil) (Figure 18). As discussed with food, bedding should be stored properly in an area that is clean, uncluttered and regularly sanitized (Figure 19).

Typically, with solid-bottom caging, the entire cage unit is changed when the bedding is changed. With suspended wire-bottom caging, the pan below the cage, which may or may not contain bedding, is changed more frequently that the actual cage itself. Water bottles and feed hoppers should be changed at a frequency that provides animals with fresh clean water and food at all times.

The frequency of cage sanitation is dependent on the frequency of bedding changes, the type of cage used (solid-bottom versus wire bottom), and the density of animals within the cage.

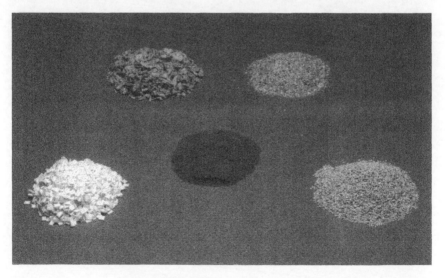

Fig. 18. A variety of contact bedding materials are used for hamsters and gerbils including (clockwise from bottom left): manufactured cellulose cubes, wood pulp, hardwood chips, corncob, and sand (center).

Fig. 19. Bedding, like feed, should be stored off the floor in a room that is on a regular sanitation schedule.

Usually the primary enclosure and accessories (wire lids, water bottles, cage rack, and automatic watering system) should be sanitized at least every 14 days. The veterinarian or Institutional Animal Care and Use Committee (IACUC) should approve exceptions to this frequency.

Although professional judgment is required in establishing the frequency of cage/bedding and water bottle changes, a typical schedule would include 1 to 2 cage changes of each per week when solid-bottom cages are used. This frequency may increase or decrease, depending on the number of animals housed in a primary enclosure, the size of the primary enclosure, or other reasons, such as not wanting to disturb a female near parturition. For animals housed in wire-bottom cages, urine and feces drop through to the litter pan below the cage, and water is often supplied through an automatic watering system. Under these conditions, the noncontact bedding is often changed 1 to 3 times per week, again depending on the housing density of the cage. The cages and racks are usually changed every 14 days.

Special Circumstances

During prepartum, postpartum, and breeding periods, frequent cage and bedding changes may be contraindicated. Pheromones necessary for breeding may be removed during the cleaning process and alter breeding results.[27] Similarly, disturbing the female during pre- and postpartum intervals may disturb her to a point where she may injure her pups.

Methods of Cage Sanitation

Cage sanitation can be broken down into 3 or 4 phases, and it is helpful if the workflow and cage wash are designed or organized to complement these phases.

Phase 1:   Cage disassembly and prewash, removes gross contamination from the cage. Solid-bottom cages are dumped and bedding that is adhering to the cage is scraped free. A bedding dump station should be used to minimize exposure of the personnel to the soiled bedding. Caging prewash, may be used to pretreat cages prior to washing. For example, cages may be hosed down prior to washing. Those cages with a urine mineral deposit may be treated with a commercially available acidic detergent to soak and remove the scale prior to washing. Automatic watering systems are also fulshed at this time.

Phase 2:   Cage washing is used to disinfect the cage. This may be done with hot water, chemical disinfectants, or a combination of the two, provided the conditions for temperature and time will kill many the organisms on the caging. The recommended water temperature for washing cages is 143 to 180°F.[27] Although hot water may be used alone, the most common method for cage cleaning is to combine hot water with a detergent to disinfect the cage. Detergents, which can withstand high temperatures, must be specifically requested. It is important to note that water bottles, sipper tubes, feeders, and other accessories must be washed using the same conditions. Likewise, when automatic watering systems are used, the watering system should be disinfected during the washing period. A hyperchlorinated flush is often used for this purpose, after the cage rack itself has been washed.

Phase 3: Cage reassembly, takes place after the cage and accessories have been washed. Clean fresh bedding is placed into the solid-bottom cage, or in the litter pan below the wire-bottom cage, the lids and feeders (where appropriate) are placed on the cage, and the cages are ready to return to service. If phases 1 and 2 are performed correctly, the cages have been disinfected when they enter phase 3 of the cleaning cycle.

Phase 4: Cage sterilization, is used in instances where pathologic organisms are known to be present, in animals with strictly regulated microbiological flora, or when sterile supplies are required for entry into the animal room (e.g., immunocompromised animals). An autoclave is used to sterilize caging.

Hand cage cleaning may be used throughout all phases to disinfect cages. The primary concerns when using hand cleaning are that you must be sure to clean all areas of the cage, rinse residual detergent/disinfectants from the cage, and take precautions to protect yourself from exposure to the chemicals and hot water used. Usually hand cleaning involves scrubbing the cage with brushes, or using a high-pressure system to spray the caging with heated water (water and detergent, followed by a water rinse) or steam.

Automated cage cleaning is a more reliable and consistent method for washing caging materials and accessories. The cage rack washer (see Figure 20) can be programmed to dispense the appropriate detergents and disinfectants during the wash cycle (e.g., acid detergents to remove urine scale), and the cycle itself can be programmed to control the length of time spent in a cage prewash, wash and final rinse. In addition, automated cage washing provides a consistent wash cycle, and can be programmed to guarantee that the wash cycle meets the 143 to 180°F requirement. After washing the shelf rack of cages, the automatic watering system manifold should be flushed out with a chlorinated rack manifold flushing system (see Figure 19 and 21). Recoil hoses used in automatic watering systems must also be disinfected on a regular basis, usually every 14 days. A chlorinated recoil hose flushing station similar to those used to flush the rack water manifold is very effective in disinfecting the recoil hose.

Fig. 20. A modern cage wash facility will often contain a automatic bottle washer (left) and tunnel rack washer (right).

Fig. 21. The rack manifold flushing station is used to flush the automatic watering system with hyperchlorinated water after the rack exits from the tunnel washing machine.

animal room sanitation

The animal room cleaning utensils (mop, dustpan and broom) should be assigned to an area and not be transported between areas of different contamination risks.[27] Ideally, each animal room should have it's own cleaning utensils. Animal room sanitation may be performed at several different intervals, depending on the species of animal housed in the room, the population density and disease status of the animals within the room, the dress code for entering the room, the types of procedures permitted in the room and the traffic within the room. A typical schedule may include daily sweeping, daily or weekly mopping (or more often as necessary) with a detergent disinfectant solution, and a scheduled room breakdown for thorough cleaning and disinfection monthly or quarterly. The daily sweeping is designed to remove gross debris that have accumulated on the floor (see Figure 22). Mopping with a detergent disinfectant is intended to remove debris, which may become ground into

Fig. 22. Animal rooms should be cleaned daily. Personnel should wear protective apparel that will prevent exposure to the cleaning agents.

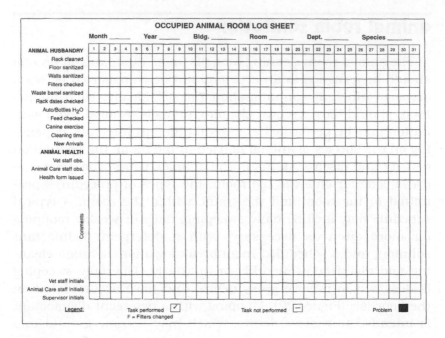

Fig. 23. A check-off log is useful to have on the door of each animal room. This log allows you to see at a glance if all procedures within the room have been completed, as well as providing documentation of what has occurred in the room.

the floor, and to control the level of bacteria present on the floor. It is good practice to autoclave mop heads after each use, otherwise the mop head can become the source to harbor microorganisms from one day to the next. Room break down is designed to periodically disassemble the entire room, disinfect floors, walls and ceilings, and perform any repairs that may not be possible at other times.

In order to keep track of sanitation procedures, it may be helpful to create a room checkoff sheet and post it outside each animal room. This form will serve as a reminder of the procedures to be followed, and document those that were performed on a daily basis (see Figure 23).

sanitation quality control

The practices employed within your facility should be monitored to ensure that your sanitation procedures are achieving the desired effect, and that equipment is operating properly (see Figure 24).

Animal room and cage surfaces, water bottles, sipper tubes, and other accessories can be cultured and evaluated for micro-biological contamination. Cultures of flat surfaces using RODAC plates (see Figure 24), and cultures of air samples, water samples, and swabs of hard-to-reach places (inside sipper tubes, corners of cages, ceiling and wall junctures) within the room or on the cage may be used to identify the sanitation levels that are present before and after cleaning. These results will also assist in establishing sanitation frequency. Likewise, cage wash-

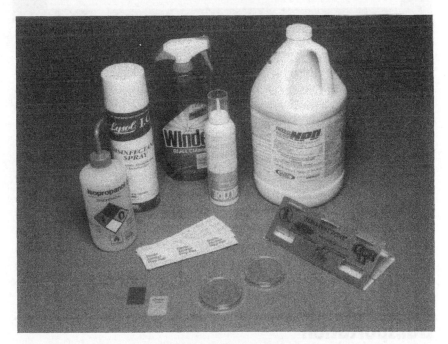

Fig. 24. A variety of sanitation supplies are used to sanitize the animal room. In addition, the room is often monitored with glue traps for the presence of vermin. Bacterial cultures may be taken of the room surfaces after sanitizing to ensure that standards are achieved.

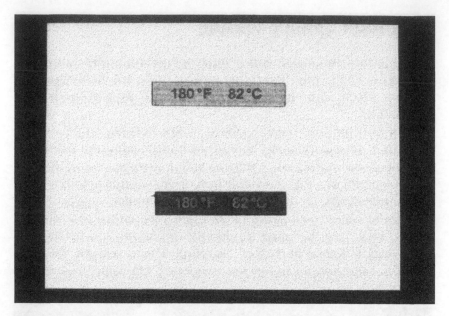

Fig. 25. Heat-sensitive temperature strips are attached to cage surfaces as a quality assurance check of the washing procedures. They will allow you to determine at a glance if the cage surface has reached 180°F during the cleaning cycle. The strip turns black (bottom strip) when the correct temperature is reached.

ing equipment can be monitored with temperature indicator strips (see Figure 25) or temperature chart recorders to provide documentation that the wash temperature has been achieved. Automated soap dispensers may be used to determine that the detergent disinfectant is being added to the wash cycle properly. Temperature guarantees can be designed into the cage wash cycle, ensuring that the cagewasher will not advance through the cycle if a minimum temperature is not reached.

transportation

When hamsters and gerbils must be transported between facilities or institutions, it is critical to ensure that they are adequately packed and protected during the process. In the United States, specific requirements for transportation of hamsters are outlined in the Animal Welfare Act regulations,[28] and should be reviewed; see Table 7). Since there are no specific requirements for transportation of gerbils, those outlined for the hamster should be applied.

Pre-Shipment Evaluation. All animals that are scheduled for shipment should be evaluated by a trained technician or veterinarian to ensure that they are in good health and will survive the shipping procedure without minimal risk to their well being. The pre-shipment evaluation may also include completing a health evaluation of the animal room, including screening sentinels for serologic or microbiologic evidence of diseases. Results of the most recent evaluation of the colony of animals should be provided to the recipient prior to shipping the animls.

TABLE 7.   SHIPPING STANDARDS FOR HAMSTERS[a]

Age	Dwarf, in^2	Other, in^2
Weaning to 5 weeks	5.0	70
5 to 10 weeks	7.5	11
Over 10	9.0	15

[a] Regulations of the Animal Welfare Act.[28]

Shipping Containers. The shipping container (Figure 26) used to transport a live hamster or gerbil should be clean, and preferably sterile. The shipping container should be constructed in a manner that prevents escape, is free from protrusions or sharp edges that can injure the animal, and should withstand stacking during transportation. If necessary, the inner surface should be covered with a fine wire mesh screen to prevent the animals from gnawing through or escaping. Ventilation openings and projections on the outside which create a space between stacked containers and allow for air flow should comply with the specific requirements outlined in the Animal Welfare Act Regulations.[28]

The interior height of the primary enclosure of the hamster transport box must be at least 6 inches, except for dwarf hamsters, in which case it must be at least 5 inches high.[28] The minimum amount of floor space for hamsters is:

Fig. 26.   Filtered shipping containers are used to transport animals. Processed or natural foods may be added to the container. Natural foods have an advantage over processed foods in that they provide animals with a source of both nutrition and water while they are in transit. Water may also be provided with water packs and through the use of gel packs.

Gerbils in transport should be provided with a primary enclosure floor space recommended above for housing animals, including an interior height of 6 inches.

Food and Water. Depending on how long the animals are to be transported, the need for food and water may vary. Animals which are transported for a period greater than 6 hours must be provided with food and water.[28] When this is necessary, the quantity and quality of food and water must be sufficient to satisfy the animals needs throughout the duration of transport. Vegetables are often given to animals to provide a single source for both nutrition and water. Standard rodent chow mixed with water, or shipping containers containing water dispensers or water gel packs are also commonly used.

Observation. During surface transportation, hamsters must be observed at least every 4 hours to assure that they are receiving adequate ventilation, their ambient temperature is within the prescribed limits, and that they are not in any obvious distress. Any finding that results in stress to the animal must be resolved. Animals that develop a health problem must be treated. Shipping containers with observation windows built into them are useful for this purpose.

When hamsters are air-shipped, they must be observed at least every four hours if the cargo space is accessible. Otherwise, they must be observed when they are loaded and unloaded.[28]

Compatibility. If several animals are transported in the same container they should be of the same species and compatible. Compatible groups should be established prior to shipping the animals. This is particularly true for adult hamsters and gerbils, otherwise they will fight if they are not familiar with their cage mates.

Environment. The environment to which the animal is exposed during shipping must be maintained carefully. Space between boxes must be provided to allow for good ventilation when stacking of boxes occurs. Large fluctuations in temperature, and temperature extremes must be avoided. Hamsters must be shipped in an environment that is maintained at 45 to 85°F unless the animals have been acclimated to lower temperatures. It is important to remember that since hamsters are natural hibernators, entering hibernation at ~41°F.[31] they more easily adapt to cold temperatures than to warm temperatures.

animal receipt

Since hamsters and gerbils may contract a variety of diseases that could alter the validity of research results, the purchase of only specific pathogen-free (SPF) hamsters and gerbils is strongly encouraged. Vendors should be asked to supply recently obtained information regarding the health status of their hamster and gerbil colonies prior to your purchase. Upon receipt, the shipping containers should be disinfected (use a detergent disinfectant such as a quaternary ammonium a chlorine based liquid misted on the outer portions of the shipping container) prior to unpacking the animals (see Figure 27). Extreme care must be used so that the animals do not come in contact with the disinfectant. Once the box has been sprayed, and sufficient time has passed for agent to disinfect the container (typically 10 to 15 minutes), the animals should then be assessed to assure that the animals meet the specification of the order (e.g., they are the correct age, weight, sex, stock or strain).

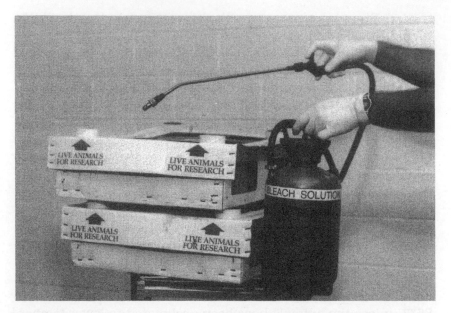

Fig. 27.   Shipping containers should be disinfected at the time of receipt. This will decrease the likelihood of exposing animals to unknown contaminants during unpacking. A fresh mixture of 10% bleach solution is often used to disinfect the outside of the container.

record keeping

The availability of records on animals within the facility, room census, and work records for the facility itself are a key component to animal husbandry. Identification of original supplier of the animal, order requisition number, breeding area of origin at the vendor, the birth date, sex and age of the animal, the protocol to which the animal is assigned, and the investigators name minimize the chance of an individual using an animal that was not assigned to them (see Figure 28). For hamsters, the easiest way to accomplish this level of identification is to create a cage card and appropriately label it with the above information.

With respect to individual identification of animals within a cage, ear tags, implantable (subcutaneous) electronic microchips, and temporary identification numbers are often used. The facility veterinarian or manager should be consulted on where

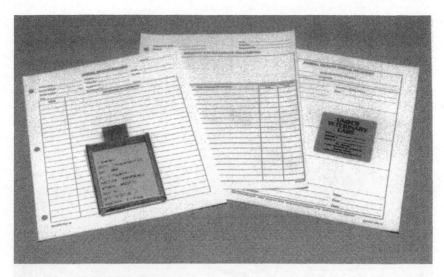

Fig. 28. Record keeping is an important factor in caring for animals. The cage card on the left contains important information on the source, age, sex, and delivery of the animals within the cage. The medical forms shown are used to document procedures and treatments that have occurred.

to obtain these devices and how to use them. Some forms and records used within the animal facility are described below.

Health Records

Any medical problem encountered with an animal should be reported in a timely manner to the veterinary staff. Once reported, health records should be maintained on the animal or group of animals receiving medical treatment. The records should include the animal's identification, a description of the medical problem, diagnostic procedures, treatment(s) and the outcome. The possibility of the medical problem being related to experimental treatments must always be considered, and any proposed treatment should be discussed with the investigator prior to proceeding.

Census

Room census should be taken at specific intervals (typically monthly) in order manage the husbandry aspects of the animal facility. The staffing required for maintaining a room, the amount of food and bedding required for the animals, the frequency of

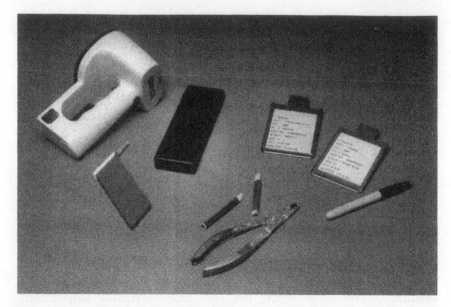

Fig. 29. Animals may be identified with both temporary and permanent methods which may include (clockwise) ear tags (bottom center), subcutaneous transponder implants (they require one of the readers shown in the upper left), a cage card, and a permanent marker (indelible ink).

room break down and cleaning, and required caging depends on the room census. Room census cards are often stored within the animal room or in the animal facility supervisor's office.

Work Records

An animal room should be looked at as one would any other piece of equipment within the facility. In order to know the condition and history of an animal room, complete records should be maintained on the room husbandry procedures that occur in the room. In order to keep track of room upkeep and sanitation procedures, it is helpful to create a room checkoff sheet and post it outside each animal room (See Figure 23). This form will serve as a reminder of the procedures to be followed, and document those that were performed on a daily basis (e.g., animal health checks, feeding, watering, temperature and humidity checks) through a check off system and initials of the person conducting the work. It also provides an excellent historical record for events that occur within the animal room.

breeding

Normative reproductive values of hamsters and gerbils are listed in Table 8. High-quality hamsters and gerbils are available from commercial breeders (see Chapter 6), or they can be bred without much difficulty.

TABLE 8. REPRODUCTIVE VALUES OF THE HAMSTER AND GERBIL

Parameters	Syrian Hamster	Gerbil[a]
Breeding age/weight (F)	56 days/90–110 g	65–85 days
Breeding age/weight (M)	63 days/100–120 g	75–85 days
Estrous cycle (days)	4, polyestrous	4–6, 4 ave., polyestrous
Duration of estrus (hours)	4–23, 20 ave.	12–18
Copulation	1 hr after nightfall	At night
Postpartum return to estrus	5–10 minutes	Immediate
Gestation (days)	16	24–26
Weaning age	21 days, 35-40 g	21–28 days, 12–18 g
Begin dry food (days)	7–9	by 16
Litter size	4-12, 7–8 ave	1–12, 5 ave.
Reproductive life, (F)	6 months–2 years	7–20 months
Reproductive life, (M)	9–12 months	24+ months
Mating scheme	Pairs or 1M to 2–4F	1:1 to 1M:3F, permanent pairing

a Breeding information and values are presented as an average or a range from data given in References 10, 13, 29, 31, 34, and 35.

Hamsters

Sexing of hamsters is relatively easy (Figures 30 and 31). The ano-genital distance is posterior due to protrusion of the testes into the scrotal sac. The female posterior is not as rounded, and the genital papillae is more prominent than in the male.

- The male is sexually mature at around 7 weeks of age and the female at 6 weeks of age.

- The optimal 24-hour light:dark cycle for successful breeding is 14:10, respectively.

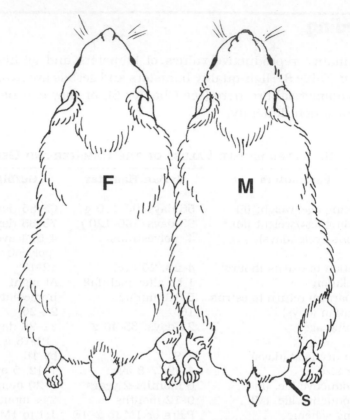

Fig. 30. Sexing hamsters may be done from above by observing the contour of the hind quarters. The male (M) has a rounded posterior which is created by the scrotal sac (S).

- Monogamous pairing is often set up soon after weaning to avoid the aggressive response females often show towards males in hand mating situations. In hand breeding systems, where the male is only placed with a potentially receptive female long enough to mate, an unreceptive female will attack and possibly kill a newly introduced male. Harem systems of 1M:4F may be used for production colonies.

- The female has a 4-day estrous cycle. They attract males through scent marking the day before they are receptive. A female will exhibit lordosis when receptive to breeding.

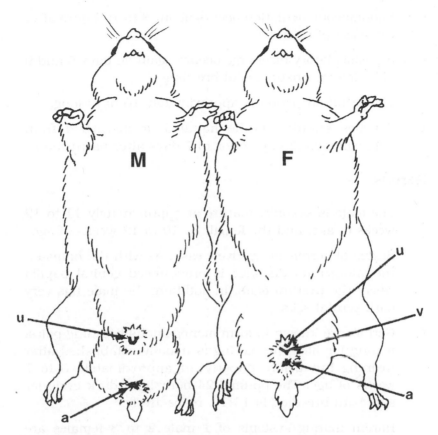

Fig. 31. When sexing hamsters from the ventral aspect, note that the distance between the anus (a) and genital papilla (u) is up to 2 times greater in the male (M) than the female (F), and the genital papilla is more prominent in the female than male. Also, note that the female has a vaginal opening (v).

Breeding will occur repeatedly for up to 30 minutes, usually within 1 hour into the dark cycle.

• The different stages of the estrous cycle are characterized by different vaginal discharges. On day 2, the postovulatory discharge is copious, viscous, thick creamy white with a distinct pungent odor. On day 3 the discharge is waxy, and on day 4 it is usually a translucent mucous discharge. Females are usually receptive to breeding 3 days after the postovulatory discharge.

- Spontaneous ovulation occurs about 8 to 10 hours after the onset of estrus.

- A postovulatory discharge occurs again on days 5 and 9 following an unsuccessful breeding.

- The gestation period is approximately 16 days long.

- Females should not be handled or disturbed from approximately 2 days prior to 7 days after parturition.

Gerbils

- The male is sexually mature at approximately 11 to 12 weeks of age, and the female at 10 to 12 weeks of age.

- Sexing of gerbils is relatively easy. As with the hamster, the ano-genital distance is pronounced genital papilla where the urethra exits. In addition, the male has very dark scrotal sacs.

- Gerbils are known to form monogamous breeding pairs, mating for life. The pairing is usually established after weaning but before puberty, at approximately 6 to 7 weeks of age. The optimal 24-hour light:dark cycle for succesful breeding is 14:10, respectively.

- Harem mating systems of 1 male:2 to 3 females are frequently used in production colonies. When this system is used, non-parturient females may cannibalize the young.

- If one mate dies, the remaining mate will often reject the introduction of a new mate.

- Even though reports indicate that a female may be introduced and receptive to a strange male during estrus,[23] timed matings (forced mating during estrus) have not proven to be successful in production colonies of gerbils.

- The estrous cycle ranges from 4 to 6 days in duration. Ovulation occurs spontaneously and the female is sexually receptive for approximately 12 to 15 hours.

- Mating usually occurs at night.

- Males will thump their hind feet during courtship, and will mate the female several times.

- Both the male and female will build a nest and care for the young. They will build a nest with cotton fiber nesting material available from bedding vendors.

- The gestation period is 24 to 26 days in duration.

notes

management

regulatory agencies and compliance

Regulations, policies, and guidelines outlining the use of laboratory animals for research, teaching and testing within the United States are summarized below. In addition to those noted below, specific regulatory agencies and their requirements may vary with the locale.

The United States Department of Agriculture (USDA)

The Animal Welfare Act contains provisions[28] to prevent the sale or use of animals that have been stolen; prohibit animal fighting ventures; and ensure that animals used in research, for exhibition, or as pets receive humane care and treatment. The law provides for regulating the transport, purchase, sale, housing, care, handling, and treatment of such animals.

The Improved Standards for Laboratory Animals Act was enacted into law as a part of the 1985 Farm Bill.[36] This new law amended the Animal Welfare Act to require new standards for the care, treatment, and use of laboratory animals, and to establish an information service at the National Agricultural Library. It also stipulated that the U.S. Department of Agriculture (USDA) must inspect each research facility, and that each facility must provide reports that verify compliance with the new regulations, provide proof that personnel involved with animal care and use are trained, and establish at least one institutional animal committee to conduct semiannual reviews.

* Specific Standards or requirements of the Animal Welfare Act are described in the Regulations of the Animal Welfare Act.[28]

* Registration with the USDA and adherence to the USDA Regulations are required by all institutions, except elementary or secondary schools using animals such as hamsters in teaching, testing or research in the United States.

* Unannounced on-site inspections by USDA inspectors employed by the Animal Care section of the USDA should be expected at least once per year. More frequent inspections are not uncommon.

* The USDA requires an annual report on animal usage covering the federal fiscal year from October 1 to September 31. The report is due by December 1.

The National Institutes of Health, Public Health Service (PHS)

Oversight responsibility is described in the Health Research Extension Act of 1985.[37] The Policy is described in the Public Health Service Policy on Humane Care and Use of Laboratory Animals.

The Policy is administered through the Office for Protection from Research Risks, which requires that an assurance be on file and approved. The Policy on Humane Care and Use of Laboratory Animals was revised in 1996, with the mandated changes incorporated into the Health Research Extension Act, Public Law 99–158. Institutions must comply with this policy if they are awarded a federal grant or contract to conduct research involving the use of vertebrate animals.

Principles for implementation of PHS policy are those described in the **Guide for the Care and Use of Laboratory Animals**.[27] These guidelines provide information on the care and use of laboratory animals in research. The recommendations in the Guide cover physical construction of animal facilities, husbandry, veterinary care, sanitation and the qualifications of personnel along with other aspects of animal care. The Guide emphasizes the role of the Institutional Animal Care and Use Committee in oversight of animal care facilities, procedures, and compliance.

The Association for Assessment and Accreditation of Laboratory Animal Care, International (AAALAC, Int'l.) uses the criteria set forth in the Guide to evaluate and accredit animal care and use programs. AAALAC, Int. accreditation is one of the highest standards achievable for documenting excellence in animal care and use.

The United States Food and Drug Administration (FDA) and the Environmental Protection Agency (EPA).

Policies are described in the Good Laboratory Practices for Nonclinical Laboratory Studies (GLP) Regulations (CFR 21 (Food and Drugs), Part 58. Subparts A–K: CFR Title 40 (Environmental Protection Agency), Part 160, Subparts A–J: CFR Title 40 (Protection of Environment), Part 792, Subparts A–L).

GLP regulations were enacted in 1979 to assure the outstanding quality and integrity of animal safety data in non-clinical laboratory studies that support or are intended to support applications for research or marketing permits. This includes studies for products regulated by the Food and Drug Administration, including food and color additives, animal food additives, human and animal drugs, medical devices for human use, biological products, and electronic products.

GLP regulations encompass: (1) organization and personnel; (2) facilities and equipment; (3) facilities operation (including animal care); (4) test and control articles; (5) protocols for and conduct of a non-clinical laboratory study; and (6) records and reports. In general, standard operating procedures must be outlined and rigorously followed and supported with detailed records.

Association for Assessment and Accreditation of Laboratory Animal Care International, Inc. (AAALAC International)

AAALAC International is a nonprofit organization designed to provide peer review-based accreditation to animal research facilities. Basis for accreditation by AAALAC International is adherence to principles outlined in the Guide. Accreditation is voluntary with on-site assessment of the animal care and use program occurring every 3 years. Annual reports on program changes are required. AAALAC accreditation is one of the highest standards achievable for documenting excellence in animal care and use.

Institutional Animal Care and Use Committee (IACUC)

The basis for achieving a high-quality, comprehensive program for animal care and use most often comes about through the close interaction of the IACUC with the department responsible for carrying out the program, and the individuals governed by the program.

Although gerbils are not a named species in the USDA Animal Welfare Act, and thus not "covered" under the regulations, hamsters are "covered" by the USDA regulations. Furthermore, a progressive program for animal care and use **will not differentiate between** "covered" and "non-covered" species of animals. Therefore, for the purposes of this discussion, the USDA, PHS and AAALAC require an IACUC at any institution using either/or hamsters and gerbils in research, teaching and testing.

Important points to note about the IACUC are:

- The Institutional Animal Care and Use Committee (IACUC) will be appointed by the Chief Executive Officer of the institution.

- The IACUC will have no less than three members to comply with the USDA regulations, and five members to comply with PHS policy.

The IACUC membership should have the following composition:

- A Chairperson

- At least one Doctor of Veterinary Medicine who has training and experience in laboratory animal medicine or science, and responsibility for activities involving animals used in the research efforts of the institution.

- One individual who is not affiliated with the institution in any way other than as a member of the IACUC. Clergy or lawyers often fill this role.

In addition, the PHS requires the following members to be on the IACUC:

• A practicing scientist experienced in research involving animals

• One member whose primary concerns are in a non-scientific area. This individual may be an employee of the institution served by the IACUC. A lawyer or a member of the Public Policy and Affairs department often fills this position.

• It is acceptable for a single individual to fill more than one of the above categories.

Responsibilities of the IACUC. The written regulations should be consulted for an in-depth coverage of the IACUC responsibilities. In general, the IACUC is charged with the following:

• Review and approve, require modifications to secure approval, or withhold approval of those components of proposed research activities involving the use of animals in research, teaching and testing. **Protocols must be approved prior to the use of the animals.**

• Once a protocol has been approved, the IACUC should review and approve, require modifications to secure approval, or withhold approval of proposed **significant changes** regarding the care and use of animals in ongoing activities.

• Review, at least once every six months (semi-annually), the research facilitie's program for humane care and use of animals to assure that they meet the standard of the regulations. This review will include an inspection of all of the research facilities, animal facilities, including animal study areas.

• Prepare reports of the semi-annual evaluations of the animal facilities and animal care and use programs and submit the reports to the Institutional Official of the research facility. The reports shall be reviewed and signed by a majority of the IACUC members and must include any minority views. The reports must distinguish significant deficiencies from minor deficiencies. A significant deficiency is one that in the judgment of the IACUC and the Institutional Official is or may be a threat to the health or safety of the animals.

- Submit written annual reports to Office for Protection from Research Risks updating the institution's assurance. These reports must include minority views filed by members of the committee.

- Assure that personnel are adequately trained and qualified to conduct research using animals.

- Assure that animals are properly handled and cared for.

- Assure that the investigator has considered alternatives to potentially painful or distressful procedures, and that the research does not include unnecessary duplication.

- Assure that sedatives, analgesics and tranquilizers are used to relieve pain and distress whenever appropriate.

- Assure that proper surgical preparation and technique are utilized.

- Assure that animals are euthanatized properly, and that the personnel performing euthanasia are properly trained.

- Review, and, if warranted, investigate concerns involving the care and use of animals at the research facility resulting from public complaints received and from reports of noncompliance received from laboratory or research facility personnel or employees.

- Make recommendations to the Institutional Official regarding any aspect of the research facility's animal program, facilities, or personnel training.

- Be authorized to suspend an activity involving animals.

occupational health and zoonotic diseases

Animals present health risks that workers should be made aware of.[38,39] Zoonoses refers to diseases that are transferred directly from an animal to a human, where the animal acts as the intermediate host or vector.

This section reviews selected known or potential zoonotic agents that have been documented in some individuals who worked with hamsters and gerbils. In no way does this chapter include all infections or diseases that are zoonotic. The selection included has been made from those that are of principal interest, for various reasons, in public health. The number of listed zoonoses increases with our expansion of biomedical knowledge, improved diagnostic competency, and improved health services. This section of the chapter simply serves an introduction to the topic of zoonotic diseases.

Hamsters and gerbils purchased from reputable vendors present a very low risk with respect to zoonotic diseases. Nevertheless, individuals who are exposed to laboratory animals must understand that however low the potential, precautions must be taken to prevent exposure, and provisions must be available to treat an individual upon exposure. Following good principles for hygiene such as wearing respiratory protection to minimize exposure to allergens, gloves when handling animals, a lab coat in the animal facility and washing your hands after working with animals are important preventative measures (see Figure 32).

In addition to the above, an occupational safety and health program should take into consideration the following possible health hazards to those individuals working with hamsters and gerbils.

Bacteria
Leptospirosis is caused by the bacteria *Leptospira spp.* Although the common reservoir for this organism includes gerbils and hamsters, this disease is very unlikely to be present in purpose-bred animals. When infected, the disease in humans

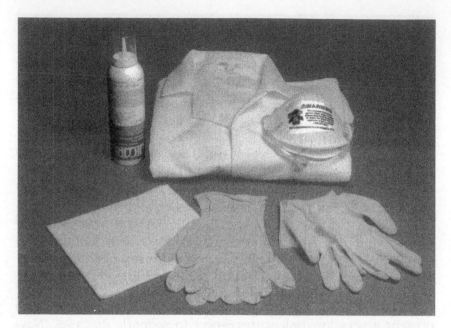

Fig. 32. Personal protective supplies commonly available in an animal facility include a lab coat, face mask, gloves, hand towels, and cleaning agents. Safety glasses (not shown) are also commonly used.

may vary from inapparent to death. Individuals may experience a sudden onset of fever and headache, leukocytosis, chills, encephalitis, retroorbital pain, conjunctival suffusion, jaundice and hemorrhage.[40] Transmission is most likely to occur through direct contact with the animal or its contaminated urine, or aerosol exposure during cage changing and cleaning. Since this is an organism primarily of wild rodents, prevention and control include wearing protective clothing, controlling the entry of wild rodents into the animal facility, use of purpose-bred animals, and vaccination of animals where appropriate.

Salmonella enteritidis* and *Salmonella typhimurium have been reported to cause enteritis in gerbils. Because people may contract *Salmonella* from gerbils, the animal colony should be screened and free of this organism. Personnel should wear personal protective clothing and follow good hygiene when working with gerbils.[40,41]

Other Occupational Health Issues

Allergies. Although allergies are not a zoonotic disease, allergies in personnel exposed to rodents are not uncommon.[38,39] The primary allergens with rats and mice are either the salivary and urinary proteins, or both.[38,39] In gerbils and hamsters, the allergens have not been identified. Personnel with allergies may experience symptoms such as rhinitis, sneezing, or redness, swelling and itching on their skin where they have been exposed to urinary proteins. In some cases, allergies to laboratory animals may progress to asthma. These symptoms may interfere with the person's ability to perform their job, and represent a serious occupational hazard that may eventually result in reassignment to another work area. Therefore, it is advised that personal protective clothing such as respiratory protection (dust and mold mask, positive-pressure HEPA filtered-air-stream helmet), gloves, and a clean launderable or disposable laboratory coat or coveralls be worn. Individuals who are known to be sensitive should also have a periodic respiratory evaluation as part of their occupational safety and health program.

Bite, puncture, and scratch wounds. Hamsters are known to bite if they are startled or irritated, and scratches may easily occur when handling hamsters and gerbils. Puncture wounds can occur when handling caging or other supplies with sharp edges. Personal protective clothing as noted above should be worn to minimize these incidents. If you are bitten, scratched, or suffer from a puncture wound, the wound should be cleaned with soap and water and you should report the incident to your facilities health-services group.

Parasites

Hymenolepis nana and *Hymenolepis diminuta*, tapeworms occasionally found in hamsters, may also infect humans. In humans, there are often no apparent clinical signs. In heavily infected individuals, nausea, anorexia, vomiting, diarrhea and central nervous signs of agitation may be seen. Control includes managing a good quality-control program with vendors, screening their colonies to ensure that they are not carrying this parasite, and control of fleas, the intermediate host for this parasite. In addition, controlling the entry of wild rodents into the animal facility will prevent exposure of research colonies. Personnel handling hamsters should be instructed to follow good hygiene.[40,41]

Giardia sp. are fairly common in hamster and gerbil colonies. Control measures include managing a good quality control program with vendors, screening their colonies to ensure that they are not carrying this protozoan. Personnel handling hamsters should be instructed to follow good hygiene. In humans, infection is often subclinical, or it may present as bloating, abdominal pain, nausea, diarrhea, and weight loss.[40,41]

Viruses

Lymphocytic choriomeningitis virus is a viral infection that can originate from hamsters, but more commonly originates from mice and transplantable mouse tumors.[42] Clinical signs in humans include mild influenza-like illness with or without central nervous system signs (headache, somnolence, paresthesia, paralysis). Transmission to humans occurs by aerosol exposure, direct contact with infected excretions, skin or mucous membranes, inhaling dust contaminated with dried excreta, hamster bites, and possibly arthropods. A quality-assurance program which includes serological surveillance of the vendors' hamster colonies, and screening murine tumors to prevent entry of the virus through transplantable tumors are the primary modes of prevention. Preventing the entry of wild rodents into the animal facility will minimize the exposure of in-house hamster colonies to the virus. In addition, all bite wounds should be decontaminated as noted above.[40]

experimental biohazards

The research studies that are conducted on hamsters and gerbils may include the use of chemical, radioactive, and infectious hazards. When this occurs, standard operation procedures in the form of a safety action plan should be developed to ensure that the hazardous materials and contaminated animals are handled safely.[38, 39]

veterinary care

veterinary supplies

The following supplies are useful for providing clinical care to hamsters and gerbils used in research, teaching, and testing. For any procedure, consult with a qualified veterinarian before performing a procedure you are not familiar with.

1. A laboratory scale for weighing the animal.

2. Disposable syringes ranging from 1 to 12 ml.

3. Disposable needles ranging from 21- to 27-gauge diameter, 1/2 to 1 1/2 in. long.

4. Blood collection tubes with no additive (for serum) and EDTA (for whole blood). Tubes requiring 0.05 to 1.0 ml are preferred since only small samples can be obtained.

5. Gauze sponges.

6. Disinfectant (povidone-iodine solution).

7. Sterile fluids such as sterile water, lactated Ringer's solution, or 0.9% saline.

8. Nail clippers.

9. Culture swabs in transport media for bacterial isolation.

10. Microhematocrit tubes.

11. Rodent restrainers.

12. An infrared tympanic thermometer for measuring body temperature.

13. A sterile pack containing gauze, cotton tip swabs, a scalpel blade-holder, sterile scalpel blades, sterile suture, tissue forceps, suture needle-holders, tissue staples and stapling device, tissue scissors, and suture scissors.

14. A small nose cone or bell jar for delivering inhalation anesthetic agents.

15. A hot-water heating pad.

Although the respiratory rate and character is normally assessed through direct observation of the animal, a pediatric stethoscope may also be used for this purpose. Additional supplies may also be used as the need arises.

physical examination

A physical examination should be performed on all hamsters and gerbils upon arrival at the facility, and on any animal that is housed within the facility that appears abnormal. Since hamsters and gerbils are often housed in pairs or groups, assessment of the cage mates should also be performed. The physical examination should be done systematically so that no area is overlooked. All findings should be recorded in a medical record for either the individual or group of animals. The physical examination should proceed as follows:

•   Observe the animal in its home cage and note any abnormal behavior. Often the first evidence that an animal may be ill is that it is isolated from its cage mates, has a hunched posture, a ruffled hair coat and is not showing normal exploratory behavior.

•   Another early indication that an animal is ill is a change in normal color, consistency, odor and/or amount of urine and feces. A fecal sample can be submitted for detection of parasites or ova, and for bacterial culturing.

- Assess the animal's food bowl and water bottle to evaluate appetite and thirst.

- A change in body weight is one of the best methods for determining if a hamster or gerbil is ill. Since hamsters and gerbils generally weigh less than 150 grams, minor changes in body weight can be a significant finding.

- When weighing the animal, examine its skin and fur for hair loss, wounds, hydration, and ectoparasites. Hydration is most easily assessed in small animals by carefully monitoring changes in body weight, and assessing the elasticity of the skin. Failure of skin over the shoulder blades to return to its normal position after it is lifted may be a sign that the animal is dehydrated.

- Palpate the animal's skin and abdomen for abnormal masses. Gently squeeze the thorax and slide your fingers caudally, allowing them to fall off the final rib and onto the abdomen. Palpate both ventrally and caudally. Be careful not to squeeze too hard, otherwise you may bruise or injure an internal organ. With your fingers, feel for firm masses. Look for symmetry in anatomy and shape when palpating the animal.

- Check the oral cavity for normal dentition. Clip overgrown teeth as they will prevent the animal from eating. When evaluating the oral cavity, carefully assess the mucous membranes for normal color. Mucous membranes should be pink. Be sure to examine the hamster's cheek pouches.

- Examine the ears for abnormal discharges or masses.

- Examine the feet and toes for lesions and clip overgrown toenails as needed.

- Eyes should be glistening and should not have any abnormal discharge or reddening of the conjunctiva.

- Assess the perineal region for fecal and urine staining and discharges from the vulva.

- Body temperature may be measured from the ear using a infrared tympanic thermometer. Normal body temperature for the hamster is 37 to 38.5°C and or the gerbil is 37 to 39°C.

- A pediatric stethoscope should be used listen to the chest for abnormal respiratory sounds.

quarantine, stabilization, and acclimation

A quarantine is used to physically separate newly received animals from the resident population until the health status of the newly received animals is assessed. In addition, a quarantine prohibits the investigator from accessing the animals to initiate studies, and it permits time for the animals to stabilize and acclimate to their new environment while recuperating from the effects of shipping. In cases where the health status of the incoming animals is not well defined by the vendor, the quarantine period may run 2 to 3 weeks.

Hamsters and gerbils may not require a quarantine period if the data available from the vendor is recent and comprehensive to provide a defined health status, and the shipping procedures do not expose the animals to any infectious agents. This is usually the case when hamsters and gerbils are purchased from a vendor with a known high quality of animals. Whatever the case, there is ample data to show that rodents need a minimum of 48 to 72 hours after shipping for physiological and nutritional stabilization before they are turned over to the investigator.[43,44]

common clinical problems

In this section, common clinical problems of hamsters and gerbils are reviewed. Many clinical problems appear similar at first, and often times the diseases are uncovered during screening while they are at the subclinical stage. If animals are purchased from a reputable vendor as pathogen-free animals, many of the diseases reviewed here may be avoided.

diseases of hamsters

The most common spontaneous diseases in hamsters are diarrhea and pneumonia.[13] Once either of these are detected, efforts are generally directed at prevention of infection and control of the disease through eradication of affected animals. Occasional clinical reports of parasitic diseases (mange or intestinal parasites) are found in the literature.

Bacterial Diseases of Hamsters

Proliferative ileitis (wet tail)

Wet tail is the most commonly recognized disease of hamsters. Wet tail occurs in 3 to 8-week-old hamsters, with a high morbidity and mortality rate. Clinical signs include diarrhea, lethargy, anorexia, fetid watery feces staining the perineum, weight loss, and sudden death within 1 to 3 days of onset. Recent findings indicate that an organism nearly indistinguishable from *Lawsonia intracellularis*, the causative agent of proliferative enteritis in swine, has been identified in hamsters.[45] The best method for eliminating this disease is eradication of affected animals. If the decision is made to treat the animals, administering tetracycline 400 mg/L in the drinking water or metronidazole at 2 mg/ml in the drinking water has also been reported to be effective. Other efforts should be directed at minimizing spread of the disease. Erythromycin and chloromycetin have also been reported to be effective. A strict quarantine and housing of animals in a filter top cage is recommended in controlling spread of the disease.[1,13,46]

Clostridium piliforme: Tyzzer's disease

Although Tyzzer's disease has only been reported occasionally in hamsters, it is possible that some cases go undiagnosed or are mistaken for wet tail. Clinically, diarrhea in the form of yellow, watery feces is the most common finding, along with anorexia and dehydration. In some cases, animals may present with sudden death with few premonitory signs. Although no treatment is reported in hamsters, other species of rodents with Tyzzer's disease have been treated with tetracycline, which would be the treatment of choice in hamsters. Since so many other rodent species are susceptible to Tyzzer's disease, control and prevention of the disease must be emphasized. This is accomplished through isolation or euthanasia of affected animals, and sanitation of the facility. Since *C. piliforme* is a spore-forming bacteria, the environment must be thoroughly decontaminated to prevent reoccurrence of the disease.[13,46]

pneumonia

Pneumonia, although rare in hamsters, is the second most common spontaneous disease reported, with enteritis being the most common.[13] *Pasteurella pneumotropica* is commonly associated with pneumonia in hamsters. Clinical signs include labored breathing, conjunctivitis, otitis interna, nasal exudate, and weight loss.[46] Chloromycetin and ampicillin have been used to treat pneumonia.[46] Carriers of *P. pneumotropica* should be eliminated.

cutaneous abscesses

Skin abscesses, often around the head, eyes and cheek pouch, related to *Staphylococcus aureus* infection due to fighting wounds have been reported. Surgical drainage and antibiotic therapy have been used. In addition, separation of the aggressive animals may be required.

salmonella

Hamsters are reported to be more susceptible to *Salmonella sp.* than other laboratory rodents.[13] Clinically animals are lethargic and have a rough hair coat, weight loss, increased respiratory rate, and distended abdomen. Efforts are directed at prevention of entry of this organism into the facility, and eradication of positive animals.[13]

Viral Diseases of Hamsters

Very few naturally occurring viral diseases outbreaks have been reported in hamsters.[13]

lymphocytic choriomeningitis virus (LCM)

Infection with LCM virus is usually clinically silent. Infected hamsters usually have chronic subclinical viuria, however, the infection may present as a chronic wasting disease. The virus affects blood vessels and the kidneys, and is shed in the urine, resulting in transmission to other hamsters and humans. Since LCM is a zoonotic disease that can cause meningitis in humans, it is important to prevent the entry of this virus into the animal colony. If detected, an eradication program should be instituted. [13,47]

sendai virus

Sendai virus is a very rare finding in hamsters. When present, it may cause fatal pneumonia. Efforts should be directed at prevention of viral introduction into the colony, and eradication of positive animals. [13,47]

Parasitic Diseases

protozoans

The protozoan *Giardia muris mesocricetus* (Figure 33) is the causative agent of giardiasis. Although often carried subclinically, large numbers of organisms have been recovered from weanlings with diarrhea. This may be an incidental finding, but

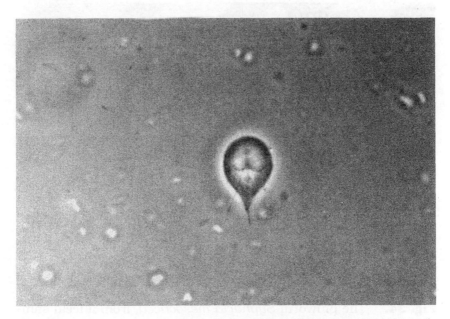

Fig. 33. The protozoan *Giardia muris mesocricetus*, found in a fecal sample from a hamster with diarrhea. (Photo courtesy of Mr. David Pavlock, Bristol-Myers Squibb).

their presence is suggestive of a causal relationship. Treatment is 0.1% dimetranidazole in the drinking water for 14 days.[48]

Spironucleus (Hexamita) muris and *Tritrichmonas sp.* have been found in the gastrointestinal tract, but few cases of clinical disease have been reported. If present, both organisms may spread to mice and rats.

nematodes

The pinworm *Syphacia muris* and *Syphacia obvelata* occur (rarely) in hamsters and *Syphacia mesocriceti* is even more rare (Figure 34). *S. obvelata* has a direct life cycle, with females laying eggs on day 11 to 15 of the life cycle.[49] Pinworms may be diagnosed by pressing a piece of clear adhesive tape against the hamsters anus, placing the tape on a microscope slide, and assessing the slide for pinworm eggs. Since *S. obvelata* and *S. muris* may spread to other laboratory rodents,[49] they should be eradicated from the hamster colony when detected. Use of an avermectin will usually eliminate the adult infestation in the

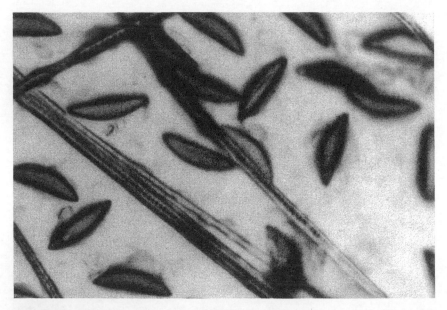

Fig. 34.   The pinworm *Syphacia mesocriceti*, from a fecal sample taken from a hamster with diarrhea. Typical of this pinworm is a flat side to the egg. (Photo courtesy of Mr. David Pavlock, Bristol-Myers Squibb.)

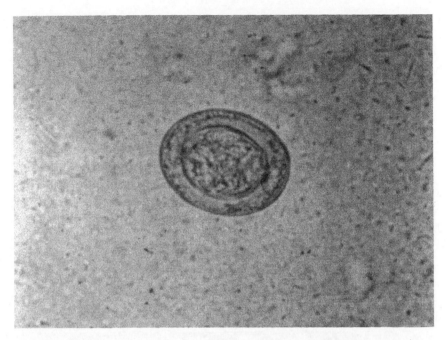

Fig. 35. This egg, found in a fecal sample from a hamster, is typical in that it is harboring the tapeworm *Hymenolepis sp.* (Photo courtesy of Mr. David Pavlock, Bristol-Myers Squibb).

animal. The environment must be thoroughly sanitized (using chlorine-based agents) to eliminate the eggs. Reinfestation with pinworms may occur due to the chemical resistant nature of the eggs. Continued monitoring in a previously positive colony is recommended.

cestodes

Hymenolepis diminuta and *Hymenolepis nana* are tapeworms that have been found in hamsters (Figure 35). Clinical signs are minimal, although intestinal impaction/occlusion has been reported with heavy infestations. *H. diminuta* has an indirect life cycle, with the intermediate host being a cockroach or grain beetle. In contrast, *H. nana* usually has a direct life cycle through the feces, but may also demonstrate an indirect life cycle through an intermediate host such as the flea or grain beetle. Since this tapeworm may be transmitted to humans, positive animals should be considered biohazardous.[13,48] Niclosamide 1 mg/gm feed (100 mg/kg body weight) for 2 weeks has been reported to cure hamsters.[48]

Ectoparasitic Arthropods

mange

Demodex criceti or *Demodex aurati* cause demodex mange (see Figure 36). *Demodex criceti* is a mite that burrows into the epidermis, feeding on the cell contents. Infestation with this mite rarely presents as a clinical disease, but may cause dry scaly skin, scabby dermatitis, and alopecia anywhere on the body.[13] *D. aurati* is only found in the pilosebaceous glands of hair follicles. Most infested animals are asymptomatic. When stressed, hair loss may occur over the back and hind quarters, but pruritus is not observed.[13,48] Treatment is not reported but methods used for other animals (cutaneous or injectable aracicide) will most likely work.

Another infestation is **itch mange**, caused by *Notedres* sp. This presents as a papular dermatitis with pruritus on the ears, nose, genital areas, tail, and limbs.[13]

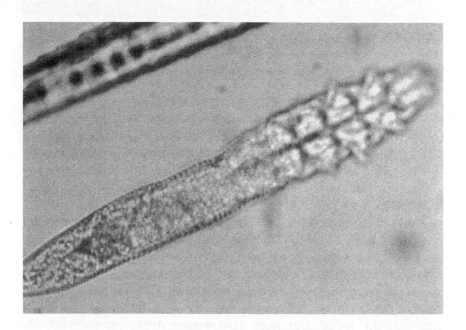

Fig. 36.   Finding this mite on a deep skin scraping of a hamster with dry scaly skin is diagnostic for demodectic mange caused by *Demodex criceti or Demodex aurati.* (Photo courtesy of Mr. David Pavlock, Bristol-Myers Squibb).

miscellaneous diseases

Amyloidosis occurs when amyloid, an immunoglobulin protein, is deposited in a variety of tissues, including the kidneys, spleen, liver, lungs and adrenals. This occurs commonly in aging hamsters and those with chronic infections. Chronic weight loss and wasting in an adult animal is the most common clinical finding.[13,50]

Atrial thrombosis with bilateral ventricular hypertrophy occurs in older animals, females more often than males, with thrombi most commonly seen in the left atria. Animals that develop thrombi may be recognized by their lethargy, subcutaneous edema, increased respiratory rate, cyanosis and increased heart rate, followed by sudden death within a week.[51,52,53]

Polycystic disease of the liver, seminal vesicles, renal pelvis, endometrium, pancreas, and seminal vesicles have been reported. The liver is the most commonly affected organ.[48,54]

common diseases of gerbils

Bacterial Diseases of Gerbils

clostridium piliforme: Tyzzer's disease

Weanling-aged gerbils are highly susceptible to Tyzzer's disease, exhibiting a high morbidity and mortality, particularly in young animals,[15] and less so in adults. Transmission occurs through contact with soiled bedding. Clinical signs include acute death, lethargy, rough hair coat, and possibly watery diarrhea. This disease should be controlled through sanitation and culling the affected animals from the colony.[13] As mentioned with hamsters, *C. piliforme* is a spore-forming bacteria that can spread to other rodents. A positive animal colony must be quarantined, and all materials used in the primary enclosure autoclaved to kill the spores. It is possible that tetracycline, which has been used to treat mice, will decrease the incidence of this disease in gerbils. Oxytetracycline added to the drinking water may also be beneficial, as is fluid therapy to dehydrated animals

nasal dermatitis or "sore nose"

This is an important disease of gerbils, particularly weanlings. Up to 20% of the animals in a colony may be affected. Erythema occurs over the dorsum of the nose, progressing to alopecia and a moist dermatitis. In extreme cases this condition may spread to the forelegs, feet, and ventral body surfaces. The etiology is unknown, but it is proposed that stress and secretion of porphyrin-containing fluid from the Harderian glands may cause irritation. It has also been proposed that this disease results from wounds sustained during burrowing. *staphylococcus aureus* is often cultured from lesions. Animals eventually recover or become unthrifty lose weight, and die.[13,55,56] Treatment consists of eliminating the factors that may cause the disease. Provision of clean, dry, soft bedding, reduction of stress, elimination of any sharp edges within the cage which may traumatize the animal, and construction of an Elizabethan collar to prevent the animal from re-traumatizing the area, will aid in treating this disease. Chloromycetin (0.083 g/1100 ml of water) and tetracycline (0.3 g/100 ml of water) for 14 days have been used to clear up infections.[57]

Salmonella sp.

Enteritis may result from bacterial infections such as *Salmonella sp.* Perianal staining, anorexia, rough hair coat, weight loss, dehydration, and sudden death sometimes occur. Some outbreaks are characterized by high mortality rates, but more typically, the animal will develop diarrhea then recover. Animals positive on bacteriological culture for *Salmonella* should be eliminated from the colony.[13,15]

Viral Diseases of Gerbils

Very little is reported on naturally occurring viral diseases in gerbils.[13]

Parasitic Diseases of Gerbils

demodectic mange

Demodex sp. has been recovered from gerbils with alopecia, hyperemia, and focal ulcerations on the legs and dorsum.[13] Mites can be seen through microscopic examination of materials obtained from a skin scraping. Treatment with a topical acaricide is recommended.

pinworms

Syphacia obvelata and *Syphacia muris* can be transmitted to gerbils from rats, mice, and other laboratory animals. The biology and treatment of pinworms is similar to that described earlier for hamsters.

cestodes

Hymenolepis diminuta and *Hymenolepis nana* occur in gerbils as well as hamsters. The biology and treatment for cestodes is similar to that described earlier for hamsters. Niclosamide is used to treat tapeworms.[15]

Miscellaneous Diseases and Conditions

epilepsy

Spontaneous cataleptic, hypnotic, or grand mal seizures related to a sudden stress (handling, noise, placement in a novel environment) have been reported in gerbil colonies. Seizure-prone and seizure-resistant strains have been described. In seizure-prone colonies, animals begin to demonstrate seizures at around 2 months of age, with 40 to 80% exhibiting seizures by 10 months of age. Daily provocation of animals will decrease the incidence through habituation of animals. Treatment with phenobarbital will prevent seizures.[13,58]

tail degloving

If a gerbil is picked up by the tip or middle of the tail, the skin over the tail may pull or slough off. This is know as degloving. Proper handling includes picking up a gerbil by the base of the tail. The tail may need to be amputated if a significant portion is affected.

dental disease

Occasionally incisors will become overgrown and require trimming, which may be done with toenail clippers or sharp cutting pliers. Spontaneous peridontal disease with plaque and calculus build-up on the teeth occurs on standard laboratory diets, but this is not considered a common clinical problem.[15]

neoplasms and cysts

Most reports of neoplasms have been in aged animals. Neoplasms of the midventral scent gland, the skin, ovary, testicles, and adrenal gland have been described. Cystic ovaries are common in the aged (greater than 2 years) female and may present as a swollen abdomen and infertility.[13,15] There is generally no treatment for cystic ovaries.

drug dosages for hamsters and gerbils

Very few drugs are specifically labeled for use in hamsters and gerbils. Administering these drugs often constitutes "extra label" use, and must be based upon sound veterinary practices. Although the dosages of drugs listed in Table 9 are from published literature, idiosyncratic responses may occur.

Note: When using these drugs, it is important to recognize that the animal's age, sex, diet, health status, metabolic rate, nutritional status, etc., can affect the results.

Antibiotics should be used after microbial sensitivity has been determined, and with knowledge of potential toxic side effects. Dihydrostreptomycin is toxic to gerbils.[59] Tetracyclines may cause fatal enterotoxemia when given subcutaneously at 50 mg/kg to hamsters, unless given simultaneously with sulfaguanidine.[29]

When administering drugs, some guidelines to injection volumes must be adhered to. Abbreviations used for the route of administration are PO for oral, IV for intravenous, IM for intramuscular, SC for subcutaneous, IN for intranasal, and IP for intraperitoneal.

Table 10 provides suggested injection volume guidelines. It should be noted that the volume given is dependent on the animal's size, the route of administration, and the rate at which the drug is given. Injection volumes for IV administration may be larger when the total injection is given slowly.

TABLE 9. DRUG DOSAGES FOR HAMSTERS AND GERBILS[a]

Therapeutic Agent	Hamster Dosage	Gerbil Dosage
Antibacterial Drugs		
Amikacin	10–20 mg/day divided q8–24h; SC, IM	10–20 mg/day divided q8–24h; SC, IM with concurrent fluid therapy
Chloramphenicol palmitate	20 mg/100 g body weight PO three times a day	—
Chloramphenicol succinate	50–200 mg/kg q8h; PO	50–200 mg/kg q8h; PO
Chloramphenicol	—	50 mg/60 ml drinking water for 2 weeks
Chlortetracycline	30–50 mg/kg q12; SC, IM	30–50 mg/kg q12; SC, IM
Ciprofloxacin	7–20 mg/kg q12h; PO	7–20 mg/kg q12h; PO
Dimetronidazole	500 mg/L drinking water	
Enrofloxacin	5–10 mg/kg q12h; PO, IM; not SC or 50–200 mg/L drinking water for 14 d	5–10 mg/kg q12h; PO, IM; not SC or 50–200 mg/L drinking water for 14 d
Gentamicin	0.5 mg/100 g body weight IM once a day	0.5 mg/100 g body weight IM once a day
Metronidazole	7.5mg/70–90g body weight q8h; PO	—
Neomycin	100 mg/kg q24h; PO	100 mg/kg q24h; PO
Oxytetracycline	16 mg/kg q24; SC	10 mg/kg q8h; PO 20 mg/kg q24; SC
Tetracycline	10–20 mg/kg q8h; PO	10–20 mg/kg q8h; PO
Tetracycline	—	250 mg/100 ml drinking water for 14 days
Tyliosin	2–8 mg/kg q12h; PO, SC, IM	2–8 mg/kg q12h; PO, SC, IM
Antfungal Drugs		
Griseofulvin	25 mg/kg q24h for 14–24 d; PO	25 mg/kg q24h for 14–24 d; PO
Ketoconazole	10–40 mg/kg/d; PO for 14 d	10–40 mg/kg/d; PO for 14 d

(continued)

TABLE 9. DRUG DOSAGES FOR HAMSTERS AND GERBILSa (CONTINUED)

Therapeutic Agent	Hamster Dosage	Gerbil Dosage
Antiparasitic Drugs		
Amitraz	1.4 ml/L topically; apply 3–6 topical treatments lightly rubbed on 14 days apart	1.4 ml/L topically; apply 3–6 topical treatments lightly rubbed on 14 days apart
Carbaryl 5% powder	Dust lightly once per week	Dust lightly once per week
Dichlorvos-impregnated resin strip	Lay 1 inch square on cage for 24h once/wk for 6 wk	Lay 1 inch square on cage for 24h once/wk for 6 wk
Dimetronidazole	0.5 mg/ml drinking water	0.5 mg/ml drinking water
Fenbendazole	20 mg/kg q24h for 5 days; PO	20 mg/kg q24h for 5 days; PO
Ivermectin	200–400 mg/kg; PO, SC; repeat in 8–10 d	200–400 mg/kg; PO, SC; repeat in 8–10 d
Niclosamide	500 mg in 150 g food or 10 mg/100 g body weight PO	1 mg/10 g body weight PO
Piperazine	3–5 mg/ml drinking water for 7 days, withhold 7 days, treat 7 additional days	20–60 mg/100 g body weight PO: use 3–5 mg/ml drinking water for 7 days, withhold 7 days, treat 7 additional days
Pyrantel pamoate	50 mg/kg; PO	50 mg/kg; PO
Sulfamethazine	1–5 mg/ml drinking water	1–5 mg/ml drinking water
Thiabendazole	100 mg/kg q24h for 5d; PO	100 mg/kg q24h for 5d; PO
Miscellaneous Drugs		
Aspirin	240 mg/kg q24h; PO	240 mg/kg q24h; PO
Atropine	0.04 mg/kg; SC, IM	0.04 mg/kg; SC, IM
Oxytocin	0.2–3 USP units/kg; IM, SC	0.2–3 USP units/kg; IM, SC
Prednisone	0.5–2 mg/kg; PO	0.5–2 mg/kg; PO
Vitamin K_1	1–10 mg/kg IM as needed	1–10 mg/kg IM as needed

a Doses are presented as an average or a range taken from data given in References 13, 15, and 60.

TABLE 10.   INJECTION VOLUMES IN MILLILITERS FOR HAMSTERS
AND GERBILS

Animal	IM	IV	IP	SC
Hamster	0.1	0.3	3–4	3–4
Gerbil	0.1	0.3	2–3	2–3

Values from Reference 29.

disease prevention through sanitation

An essential component of disease prevention is good sanitation. Hamster and gerbil cages should be changed routinely, and cleaned and disinfected as described in Chapter 2. Accumulation of excessive amounts of urine, feces, hair, and dander should be prevented. Instruments used to handle or work with animals should be routinely cleaned and disinfected. Personnel should maintain good hygiene, washing their hands with an antiseptic soap after handling animals, and wearing laundered clothing while in the animal facility. In addition, hamsters and gerbils identified as being infected with a pathogenic organism should be quarantined, treated, or eliminated from the colony. Likewise, since rats and mice carry many diseases which can be transmitted to hamsters and gerbils, efforts should be put forth to maintain pathogen-free mouse and rat colonies.

anesthesia, analgesia, and sedation

The purpose of a sedative, analgesic, or anesthetic is to minimize or alleviate the pain or distress that the animal may experience during experimental use. In order to facilitate the selection and administration of these agents, Tables 11 to 13 provide guidelines for selected dose ranges. Please note that effective doses may vary from those included in the tables due to differences in the age, strain, sex, and physical status of the animal. Because of this variation, until you become experienced with the anesthetic you select, a dose at the low end of the recommended range should be used. For additional information on the selections and use of these agents, a veterinarian should always be consulted.

TABLE 11. COMMON ANESTHETIC DRUGS FOR HAMSTERS

Agent	Dose/Route	Duration of Anesthesia (min)	Ref.
Fentanyl-Fluanisone + Diazepam	1 mg/kg IM 5 mg/kg IP	—	61
Ketamine	100 mg/kg IM	—	62
Ketamine	200 mg/kg IP	—	63
Ketamine + Acepromazine	150 mg/kg IM 5 mg/kg IM	—	61
Ketamine + Acepromazine	40 mg/kg IM – 100 mg/kg IP 1–4 mg/kg IM	—	29
Ketamine + Diazepam	40–100 mg/kg IM or IP 0.5 mg/kg IM or 5 mg/kg IP	—	29
Ketamine + Xylazine	50–100 mg/kg IP 5 mg/kg IP	28	29
Ketamine + Xylazine	200 mg/kg IP 10 mg/kg IP	—	61
Pentobarbital	50–90 mg/kg IP	30–45	65 61
Telazol + Xylazine	30 mg/kg IP 10 mg/kg IP	30	29

Anesthesia is associated with a loss of feeling or sensation. The types of anesthesia which may be useful for hamsters and gerbils can be classified into the following categories:

1. **General anesthesia** produces a state of unconsciousness, with absence of pain sensation over the entire body (e.g., methoxyfurane, isoflurane, pentobarbital).

2. **Local anesthesia** is that which is limited to a confined (small) area of the body (lidocaine).

3. **Dissociative anesthesia** is produced by interruption of information from the conscious part to the unconscious part of the brain (ketamine).

TABLE 12. COMMON ANESTHETIC DRUGS FOR GERBILS

Agent	Dose/Route	Duration of Anesthesia (min)	Ref.
Fentanyl-Fluanisone + Diazepam	0.3 mg/kg IM 5 mg/kg IP	—	61
Ketamine	40–44 mg/kg IM or IP	—	64 29
Ketamine + Acepromazine	75 mg/kg IM 3 mg/kg IM	—	61
Ketamine + Diazepam	50 mg/kg IM 5–10 mg/kg IP	—	61 29
Ketamine + Xylazine	50 mg/kg IP or IM 2 mg/kg IP or IM	—	29 61
Pentobarbital	60 mg/kg IP	30–45	64 29
Metomidate + Fentanyl	50 mg/kg SC 0.05 mg/kg SC	—	63

TABLE 13. DRUGS AND DOSES FOR SEDATION AND PREMEDICATION OF HAMSTERS AND GERBILS

Drug	Hamster	Gerbil
Acepromazine	0.5–5	Do not use, may cause seizures in gerbils.
Diazepam	3–5	3–5
Midazolam	1–2	1–2
Xylazine	8–10	5–10
Atropine	0.05	0.05
Glycopyrrolate	0.01–0.02	0.01–0.02

[a] All values are given in mg/kg IM or SC and are derived from Reference 66.

4. **Neuroleptanalgesia** is a state of quiescence, altered awareness, and analgesia with a narcotic and neuroleptic agent (fentanyl-droperidol), along with a state of unconsciousness.

principles of general anesthesia

Administering Anesthesia

When administering anesthesia, the first and foremost concern is for the animal's safety and health. This includes making sure that the animal is healthy and well-acclimated to its environment, and is of acceptable health status to be anesthetized.

There are several general rules that should be followed during the preanesthetic and anesthetic stages:

1. Fast animals 8 to 12 hours prior to anesthesia to reduce intestinal engorgement related to recent food consumption, thus decreasing the potential for misplacing an IP injection into an abdominal organ.

2. Perform a physical examination on all animals to evaluate their health status. If evidence of disease is found, elective procedures involving anesthesia are discouraged.

3. Acclimate animals to restraint devices, if they are used, in order to minimize distress and anxiety associated with their use during induction of anesthesia.

4. Accurately weigh the animal immediately before calculating the anesthetic dose.

5. If atropine is to be administered, do so within 30 minutes of anesthetic induction.

6. Accurately calculate dosages and measure doses as closely as possible. The margin of error for misdosing a hamster or gerbil is decreased due to their small size; therefore, diluting out an agent 1:5 or 1:10 with sterile water or saline allows for more accurate dose measurement, and will facilitate drug absorption if given IM, IP, and SC.

7. Assess your anesthetic technique in a pilot study if you are unfamiliar with the procedures and anesthetic effects.

8. Administer preanesthetic agents (sedatives or tranquilizers) to alleviate fear and apprehension in the animal.

9. Monitor the animal's depth of anesthesia (methods discussed later in this chapter) and respiration. Respiration should be slow and regular during surgical anesthesia.

10. Provide supplemental heat to the animal by placing the animal on a circulating warm-water blanket. This is particularly critical for smaller animals, which lose heat rapidly due to their large body surface to body mass ratio.

11. Administer supplemental fluids to animals, particularly if there is blood loss or if the surgery is longer than 1 hour and the animal will recover. Fluids may be given by the subcutaneous or intraperitoneal route. Administer at least as much fluid as blood is lost.

12. Monitor the animal's vital signs (respiratory rate, body temperature) during the recovery period to ensure that they are progressing toward consciousness. The animal should be monitored periodically until it is awake.

13. Prepare for administration of postoperative analgesics prior to initiating the procedure. In some cases, you may wish to administer the analgesic preemptively. However, this should be done with caution as analgesics will often reduce the dose of anesthetic agent required.

14. Inhalant anesthetics must be used with a scavenging system to prevent anesthetic gasses from contaminating the room.

Stages of Anesthesia

During anesthetic induction, and throughout the anesthetic period, there are four broad stages of anesthesia one must be familiar with.

Stage 1: Voluntary excitement. This stage lasts from initial administration to loss of consciousness and provides a variable degree of analgesia. Animals may become excited, vomit, urinate, and struggle if this stage is prolonged.

Stage 2: Involuntary excitement: During this stage there is development of unconsciousness and onset of regular breathing. Animals may whine, cry, and bellow, respiration is uneven, and breath-holding is common.

Stage 3: Surgical anesthesia.   The animal is unconscious and reflexes are depressed. Muscle relaxation occurs and respiration is slow and regular. The animal does not respond to noxious stimulation.

Stage 4: Medullary paralysis.   The central nervous system (CNS) is extremely depressed, respiration has ceased, and the heart beat is slowing and will eventually arrest. Anesthesia must be discontinued and the animal's respiration must be supported if the animal is expected to recover.

Assessment of Anesthetic Depth

In order to determine that the animal is in Stage 3 (surgical anesthesia) and not Stage 2 (non-surgical) or Stage 4 (medullary paralysis), it is critical to monitor the depth of anesthesia. Methods used to assess depth are as follows:

1.   **Depth assessment** — There is no single objective measurement for assessing depth of anesthesia; therefore, several indices are used to determine the animal's depth and stage of anesthesia:

    a.   Ocular mystagmus is the side-to-side movement of the eyeball during light anesthesia (Stage 2).

    b.   Palpebral (blink) reflex occurs when the corner of the eyelid is lightly touched. This reflex is usually absent during surgical anesthesia (Stage 3).

    c.   Corneal reflex results in blinking when the cornea is lightly touched. Usually present during surgical anesthesia (Stage 3) and absent during deep anesthesia (Stage 4).

    d.   Muscle tone is usually present in light anesthesia (Stage 2) and absent in surgical anesthesia (Stage 3).

    e.   Respiration is usually regular and deep during surgical anesthesia (Stage 3). Breathing may be irregular and slow during light or deep anesthesia (Stage 2 and 4, respectively). If respiration speeds up, the animal may be getting light and require more drug to deepen the anesthetic plane.

f. Responses to painful stimuli are generally present during light anesthesia (Stage 2) and absent during surgical anesthesia (Stage 3). In most animals the pedal reflex (response to a lightly pinched toe with forceps using a pressure that would be painful in an awake animal) is absent during surgical anesthesia (Stage 3). However, loss of response to painful stimuli does not occur uniformly in all areas. It may be possible to perform abdominal surgery in an animal without any evidence of response to pain, yet the animal continues to withdraw its limb when its toe is pinched. An alternative to pinching the toe in rodents is pinching the tail (remember that hamsters don't have a tail). Rodents usually do not respond to the tail pinch when they are at the stage of surgical anesthesia.

g. Most anesthetics cause a dose-dependent depression of the cardiovascular system. The respiratory system is difficlt to evaluate; however, the color of the mucous membranes (normal is red to pink; abnormal is prey to white) is a reflection of adequate oxygenation and respiration. The cardivascular system usually demonstrates a fall in blood pressure at Stages 1 or 3, which in larger animals can be indirectly monitored with a blood pressure machine.

Postanesthetic Monitoring

During the recovery period, many of the above indices can be monitored to determine the animal's progress toward recovery. In addition, provisions should be made for the following:

1. The recovery area should be warm and quiet. Supplemental heat (a bottle filled with warm water or place the cage on a circulating hot-water blanket) may be provided to aid in maintaining an animal's body temperature.

2. If possible, allow the animal to recover in it's home cage. However, do not place recovering animals in a cage with animals that have not been anesthetized or surgically manipulated.

3. The animal's vital signs (respiratory rate) should be monitored frequently until the animal is conscious. Once the animal has regained consciousness, the surgical site should be monitored frequently the first day postoperatively and at least daily thereafter.

4. During the first day, the most critical concern is that the surgical incision remains sutured and that there is no bleeding as the animal moves around. Although a slight amount of blood and serum seepage is normal after surgery, an animal should never be returned to its cage until all bleeding has been controlled. Postoperative bleeding must always be attended to immediately.

5. Analgesics should be administered as instructed by the veterinarian during the recovery period.

After 24 to 48 hours of recovery, postoperative concerns are mainly to ensure that the surgical site heals correctly and that the site remains infection-free. Appropriate wound dressings and medications should be given if evidence of infection occurs (e.g., cloudy discharge, reddening, abscess formation, purulent discharge, wound dehiscence).

Pre- and postoperative records must be maintained by the investigator or designated assistant. These records should include information regarding preoperative physical examination results, anesthesia techniques (drug, dose, and route), and evidence that the animals were monitored during the recovery period. These records should also include postoperative medications that were administered and any special care given (e.g., suture removal, wound cleaning, etc.).

characteristics of commonly used injectable anesthetics

- **Diazepam** is a benzodiazepine drug; a potent tranquilizer which produces muscle relaxation, is an anti-convulsant, and minimally depresses the cardiopulmonary systems. Diazepam is often combined with ketamine to produce balanced anesthesia. Diazepam is insoluble in water and causes muscle irritation if administered IM.

- **Midazolam** is also a benzodiazepine drug, but is approximately 3 to 4 times as potent as diazepam; but unlike diazepam, is water-soluble. Midazolam is used in combination with telazol and ketamine to produce neuroleptanesthesia or balanced anesthesia, respectively. Midazolam may cause respiratory depression, apnea, and hypotension.

- **Fentanyl** is a narcotic analgesic which also causes sedation, decreased motor activity, and an exaggerated response to loud noises (hyperacousia). Like other narcotics, fentanyl will cause hypotension, decreased cardiac output, and respiratory depression. Fentanyl is often combined with other agents such as **fluanisone** (a neuroleptic) to produce neuroleptanalgesia. By combining fluanisone with fentanyl, analgesic activity is potentiated and cardiopulmonary depression is limited.

- **Ketamine** is a dissociative anesthetic (interrupts flow of information from unconscious to conscious parts of the brain) has a short duration of action, and is rapidly metabolized by the liver. As an anesthetic, it does not produce muscle relaxation and increases salivation. It produces minor effects on blood pressure and cardiac output, mild respiratory depression, and little or no effect on pharyngeal or laryngeal muscles, thus animals retain the swallowing reflex.

- **Xylazine,** when combined with ketamine, will allow the ketamine dose to be reduced by about 1/3 of that required with ketamine alone. Xylazine has sedative, analgesic, and muscle-relaxing properties. Ketamine/xylazine cocktails will produce good muscle relaxation and a smoother recovery than with ketamine alone. Addition of xylazine will cause respiratory and cardiovascular depression.

- **Pentobarbital** is a commonly used barbiturate in which surgical anesthesia is attained at doses close to those which cause respiratory failure. Pentobarbital will stimulate hepatic microsomal enzymes. About 90% is metabolized by the liver and 60% is excreted in the urine. At anesthetic doses, it will cause hypotension, decreased cardiac output, and decreased heart rate. Respiratory depression may lead to apnea.

Principles of Gas Anesthesia

Inhalant anesthetic agents are administered best with a precision vaporizer which may require an induction chamber or tracheal intubation. Consult with your veterinarian before using these agents.

- **Halothane** is administered by inhalation with a precision vaporizer at 2.5 to 4.0% for induction and 0.5 to 2% for maintenance of anesthesia. Halothane can cause hypotension, decreased cardiac output, and respiratory depression. Halothane also sensitizes the heart to catecholamines, which may lead to an arrhythmia. Up to 20% is metabolized by the liver, with the remainder being excreted through the lungs.

> **Note:** Use of halothane, like any inhalant anesthetic, requires removal of the waste gas.

- **Methoxyflurane** is administered by inhalation using an open-drop method, or through a vaporizer at 3 to 3.5% for induction and 0.4 to 1% for maintenance of anesthesia. When using the open-drop method, gauze or an absorbent material is placed in the bottom of a bell jar and soaked with the anesthetic. A wire platform is placed over the gauze, allowing the animals to stand in the bell jar without coming in contact with the anesthetic agent.

Animals are then placed in the bell jar, the lid is closed, and they are observed until they become anesthetized. Upon becoming anesthetized they can be removed from the jar and a nose cone containing gauze soaked with the anesthetic is placed over their muzzles, allowing you to work on them.

Methoxyflurane produces a slow induction and recovery from anesthesia. Mildly decreased blood pressure and mild respiratory depression occurs.

Note: Remember that all work with anesthetic agents must be done in such a way that exposure to waste gasses is minimal.

- **Isoflurane** is administered by inhalation using a precision vaporizer at 2.5 to 5% for induction and 1.2 to 2.3% for maintenance of anesthesia. Cardiovascular depression is minimal, but respiratory depression may occur. Isoflurane is not metabolized to any great degree, and is excreted primarily through the lungs. Induction and recovery from anesthesia is very rapid.

preanesthetic management

Premedication, Sedation, Tranquilization, and Chemical Restraint

Sedatives and tranquilizers produce a mild degree of central nervous system depression, but they do not all provide analgesia. These agents calm the animal and reduce apprehension. They are most often used as a preanesthetic to reduce fear and struggling, and to decrease the amount of anesthetic needed. When used postoperatively, they will prolong recovery from anesthesia.

- **Acepromazine.** An effective tranquilizer when given IM or SC. The lower dose range is given for IM injections and the higher dose for SC injections. Because gerbils are prone to epileptiform seizures, acepromazine is not recommended as a tranquilizer or preanesthetic.

- **Diazepam and midazolam.** Not highly water-soluble and may be an irritant when given as an IM injection. However, it will produce mild sedation and is acceptable for gerbils. In contrast, midazolam is water-soluble and more appropriate for IM injections. If midazolam is substituted for diazepam, the dose is approximately 1/2 that of diazepam.

- **Xylazine.** Produces very good sedation alone, and is often combined with ketamine to provide anesthesia with good muscle relaxation.

- **Atropine and glycopyrrolate.** Both may be given preanesthetically to decrease salivation and respiratory track secretions in rodents when ketamine is administered. This effect assists in maintaining an open airway in animals during anesthesia.

aseptic surgery

Asepsis
Asepsis is the absence of microorganisms. Aseptic surgery involves preventing contact with microorganisms. Asepticism is defined as the principles and practices of aseptic surgery.

Operating Room Procedures
Most of the information contained in the following sections on operating room procedures, surgery, suture selection, and suture patterns can be found in standard veterinary surgery texts.[67,68] Consult your veterinarian for guidance in the correct selection of surgical materials. When performing surgery on a rodent, the area in which the surgery is performed must be clean, decontaminated and dedicated solely for the purpose of performing surgery; i.e., no other procedures should be performed in that area during the surgical procedure. Although a dedicated operating room is not required for rodents,[27] the trend with modern research facilities is to require that rodent surgery be performed in a dedicated rodent operating suite. When a dedicated operating room is used for the surgery, proper operating room attire must be worn at all times when in the operating room. The optimal attire for entering a rodent surgery includes wearing a surgeon's cap, face mask, scrubs, and booties. Min-

imal attire includes wearing a surgeon's cap, face mask, and booties. A face mask must be worn over the mouth and nose whenever sterile instrument packs are open or a surgical procedure is in progress.

The sole purpose of the operating room (OR) is to treat the patient. Regardless of whether the patient is human or non-human, specific principles must be established and adhered to. When a dedicated OR suite is used, the following are suggested principles for managing the OR:

1. Gowning, gloving, suturing, etc., require practice before they are used in surgery.

2. Asepsis is an absolute requirement. When in doubt, consider an item contaminated and re-sterilize it.

3. The surgeon and assistant must be completely familiar with the surgical procedure.

4. No one should enter the OR area unless they are properly attired, i.e., wearing surgical scrubs, scrub cap, special hood or face mask, and booties if required.

5. Everything that can be possibly sterilized should be sterilized if it will be used for surgery.

6. Traffic in the OR should be restricted.

7. Strict housekeeping and clean-up procedures must be followed.

Patient Preparation:

1. Prior to surgery, the patient should be thoroughly evaluated as described previously.

2. Preanesthetic and anesthetic drug doses should be calculated based upon the animal's current fasted weight.

3. Gerbils and hamsters may be fasted for 8 to 12 hours prior to abdominal surgery to decrease intestinal engorgement.

4. The potential for reduction of contamination of the operating field should be taken whenever possible. The following measures may be taken:

a. Before anesthetizing the animal, clean widely around the surgical area if the patient is cooperative. This will minimize the time the animal must spend anesthetized.

b. After the animal is anesthetized, clip away the hair from around the surgery site and clean the surgery site (to be described).

c. Administer antibiotics during surgery and/or postoperatively if needed. Never use antibiotics in place of aseptic surgical technique. However, they may be appropriate in prolonged surgical procedures.

d. Use sterile drapes to isolate the operative field, provided they do not obstruct the surgery site or the ability to monitor the animal.

5. If preanesthetics are to be used, administer them 15 to 30 minutes prior to anesthetizing the animal.

Surgical Site Preparation:

1. Clip hair away from the surgical site. Clipping widely around the proposed site and extend the clipped area if necessary.

2. Disinfect the surgery site with an antiseptic scrub (e.g., Betadine scrub) and a rinse. Typically a 70% ethanol may be used to rinse away the scrubbing soap, but in small rodents such as the hamster or gerbil, this may cause hypothermia. Sterile water may be more appropriate. Three rounds of scrubbing and rinsing is recommended when preparing the site.

3. Surgery site disinfection should begin at the intended incision line, with scrubbing being done in a circular motion, proceeding outward with each revolution. The sponge should never be brought back from a contaminated edge of the surgical area to the clean center. The sponge should be discarded after one pass over the surgical field.

4. A new, sterile sponge should be used for each cycle of scrubbing and rinsing.

5. After the site is prepared, the area is draped and the animal is positioned for surgery. A final preparation antiseptic (e.g., Betadine solution) is applied before surgery.

Surgeon Preparation:

1. High standards of personal hygiene are necessary for the surgeon and assistant to maintain aseptic technique.

2. The fingernails of the surgeon and assistant(s) should be clipped short, and all street clothing, watches, bracelets, and rings removed.

3. When a dedicated OR is used for the surgery, proper operating room attire must be worn at all times when in the OR. The optimal attire for entering a rodent surgery includes wearing a surgeon's cap, face mask, scrubs, and booties. Minimal attire includes wearing a surgeon's cap, face mask, and booties. A face mask must be worn over the mouth and nose whenever sterile instrument packs are open or a surgical procedure is in progress.

4. Hand scrubbing can be accomplished through a variety of routines. The following routine is suggested:

    a. The surgeon's cap, face mask, booties, and scrub suit must be adjusted correctly before beginning scrubbing. Hands and arms are thoroughly wet to above the elbow. Hands are held upright so that water runs down off the elbows into the sink (not onto your scrubs).

    b. A liquid soap is used to prewash the hands and arms and is rinsed off. Fingernails are cleaned (underneath) thoroughly, and a sterile brush or sponge is opened. One hand at a time is cleaned. The sponge is used to brush under the fingernails, each finger is brushed with several strokes on each of the 4 sides. As you move down the hand, each side is brushed. Once an area is cleaned the brush is never moved back up to that area because bacteria may be carried up with it.

    c. The arm is scrubbed down to the elbow. After completing one arm, allow the lather to remain on that arm while scrubbing the other arm. The brush is rinsed off and transferred to the opposite hand (clean scrubbed hand) for scrubbing. The entire process is repeated on the opposite hand and arm.

d.   Once scrubbing is completed, both arms can be rinsed off, allowing the soap to run from the finger tips off the elbows and into the sink. A final lathering and rinsing may be done on both arms if necessary. This entire process may take 7 to 10 minutes.

Gowning

If a sterile gown is worn, it should only be considered sterile from the mid-chest to the top of the surgical table. The sterile pack is opened by the OR nurse or the assistant surgeon, exposing the sterile hand towel and gown. The towel is lifted with one hand. One arm is dried and the towel is transferred to the dry hand. The opposite arm is dried with a clean area of the towel. The towel is dropped to the floor or a table top. Grasp a gown and remove it from the pack. Allow the gown to unfold but hold it high enough to keep it from touching the floor. Do not touch any part of the gown's exterior. slip the arms into the sleeves and allow the OR assistant to adjust and tie the gown from the rear.

Gloving

Closed-gloving involves picking the sterile gloves up without pushing your arms through the gown sleeve and slipping them on your hands. Open-gloving involves pushing hands through the end of your sleeves, picking up the glove (from the inside) and slipping them on. After gloving, the surgeon should always keep his or her hands above the waist. The animal can now be covered with sterile drapes.

basic surgical techniques

Surgical Principles to Promote Wound Healing

1.   Ascepsis controls postoperative infections.

2.   Gentle tissue handling minimizes tissue trauma.

3. Hemostasis minimizes blood loss, decreases infections, and preserves blood supply for the healing process.

4. Wound irrigation with sterile saline cleans the surgical site and keeps tissue moist and alive.

5. Appropriate selection of suture material minimizes wound infection and promotes healing.

6. Correct suture patterns and wound closure allow the wound to heal quicker.

7. Postoperative wound management decreases healing time.

8. A tissue at rest heals best. Try to minimize the animal's movement during recovery.

9. Removal of all sponges and instruments from the surgical site is critical. Count your sponges!

10. Wounds heal from side to side and not end to end. It is better to make the incision longer and improve exposure than to stretch a small incision.

11. Monitor the animal during recovery to ensure that it does not damage the surgery site, and that complications are attended to.

12. Sutures should be removed after 7 to 14 days of wound healing.

Wound Closure

For recovery surgeries, several tissues must be held in opposition while the healing process takes place. Tissue may be held together with sutures, staples, clips, adhesive skin strips, or surgical glue (cyanoacrylate). In order to provide skin apposition throughout the heling process, the correct shoice of wound closure material is required.

Complications of Wound Healing

Occasionally, in spite of the many precautions taken, complications arise in wound healing. This is particularly true with animals since they may chew, lick, and untie sutures. The most serious complications are postoperative infections and/or wound disruption, i.e., dehiscence. If wound infection occurs, a culture of the discharge or incision site should be taken for identification of the causative microorganism, and to determine an appropriate antibiotic regimen. If wound dehiscence occurs, the animal must be reanesthetized and the wound cleaned and reclosed.

guidelines for selecting sutures

Types of Sutures

These are two types of suture: nonabsorbable and absorbable. Nonabsorbable suture remains where it is placed, unless used to close the skin, in which case it is removed at a later date. Absorbable sutures are degraded by enzymes at various rates, eventually dissolving. A second subdivision of suture material is monofilament (single-strand) or multifilament (multi-strand). Multifilament sutures tend to have better known strength; however, they will harbor microorganisms and act as a wick for microbial entry into the surgical site. Monofilament sutures resist microbial infection, and the knot ties down smoothly.

The Size and Strength of a Suture

The size and strength of a suture varies. The size is indicated numerically and denotes the diameter. As the number of zeros (0's) increases, the size decreases. 5-0 is smaller than 4-0. The smaller the size, the less the tensile strength. When selecting a suture, the tensile strength of a suture material does not need to exceed the tensile strength of the tissue. By following this principle, use of small size sutures is promoted. However, since tissue enzymes decrease suture strength with time, loss of suture strength must always be considered when selecting a suture.

Sutures for Closure of Incisions

- **Peritoneum** — This layer is usually closed with the posterior fascia or an overlying muscle layer. A monofilament (stainless steel, proline) or multifilament suture may be used. Either absorbable or non-absorbable suture may be used.

- **Fascia** — This layer of connective tissue overlying muscle bears the maximum stress of the wound and must be closed with a moderate sized nonabsorbable or slowly absorbable suture. Monofilament (stainless steel, proline) or multifilament suture may be used.

- **Muscle** — Sutures are usually of the same materials as those used for fascia.

- **Subcutaneous fat** — Will not tolerate sutures well. However, to close dead space, fat may sometimes be sutured. Sutures used to ligate blood vessels (silk, gut) may be used.

- **Subcuticular tissue** — This tough layer of connective tissue underlies the skin. When sutured, it will hold the skin in apposition. Most frequently, if minimal tension is present, surgical gut or polymeric materials are used.

- **Skin sutures** — Monofilament sutures, staples, wound clips, and surgical grade cyanoacrylate tissue adhesive are used for skin closure. Multifilament sutures may act as a wick to draw microorganisms into the wound, so they are not recommended. When a subcuticular suture pattern is used, skin sutures are often removed by day 10.

Organ and Tissue Sutures

- **Stomach** — The stomach heals rapidly, thus, absorbable sutures are acceptable.

- **Small intestine** — Absorbable monofilament sutures are often used, as the small intestine heals quickly.

- **Colon and rectum** — Leakage of colonic contents can cause peritonitis. Therefore, a non-wicking monofilament suture that is non-absorbable or slowly absorbable should be used. The mucosal layer should not be penetrated with the suture, if possible.

- **Respiratory tract** — Monofilament sutures are the best choice to minimize the potential for infections.

- **Cardiovascular systems (vessels)** — Ligating can be done with small or medium-sized monofilament or multifilament sutures, providing that the knot strength is good. For anastomosis of vessels, a monofilament suture is used.

- **Urinary tract** — Nonabsorbable sutures must not be used as they will create a site for future infections.

Suture Needles

Suture needles come in many varieties and sizes and may be either straight or curved. The points of suture needles may be tapered (round), may have a cutting edge like a bayonet, spear, or trocar or may be blunt. Except where tissue resistance (in the skin, for example) demands a cutting point for each penetration, tapered needles are commonly used, since they produce minimal trauma. The elastic tissues soon obliterate the small, circular hole made by the round point; whereas the needles with a cutting edge leave a pathway through the tissue so that undue tension on the suture may cause it to tear. Blunt needles are used to dissect through friable or parenchymatous tissues such as the liver, kidney or spleen.[67,68]

Needles may have eyes like ordinary sewing needles; they may be eyeless and swaged (attached to the end of the suture) or they may have spring eyes (French needle). Swaged needles (attached to the end of the suture) are always used for delicate work where it is essential to minimize trauma, and are the most common type of needle used for recovery surgeries. Needles should be threaded carefully from the inside (if curved) without tension, which may cause fraying.

Suturing

Good apposition of tissues is dependent on choosing the right suture material, needle, and pattern and then using them properly. A needle is less likely to break and is more easily directed if it is held along the shank rather than near the point or the eye. Trauma can be minimized by using the smallest needle that will accomplish approximation of the tissues.

There are many varieties of suture patterns, but the beginning surgeon should attempt to master only the few basic patterns that are routinely used. The choice of pattern should be based on three criteria: (1) will it hold the tissues in the requested position without undue tension; (2) does it distribute tension on the suture material and tissue in such a manner as not to exceed the tensile strength of the suture material; and (3) does it require a minimum amount of suture material?

Surgical Knot Tying

Dexterity and speed in tying knots can be acquired only by practice, and the surgeon should practice tying knots until it becomes second nature. Knot tying should be practiced on a board covered with material, and not on the animal. The knot must be tied firmly so that it will not slip, but the tissues must not be pulled together so tightly that the blood supply is cut off and healing is impeded. Although a wide variety of complicated knots can be used, the surgeon should concentrate on using the square knot.

For rodent surgery, most knots are tied with a needle-holder, so this technique should be practiced. The needle-holder is useful when the end of the suture is short or when the suture material or gloves are slippery. The needle-holder should be applied only to the end of the sutures because the jaws of this instrument will break fibers it holds. When tying knots with your fingers, the surgeon should do so with a steady, equal pull on each end of the suture material.[67,68]

Suture Removal

Removal of exposed sutures should be done sometime between the 4th and the 14th postoperative day, depending on (1) the extent of the wound, (2) evidence of infection, and (3) the physical stress to which the wound may be subjected.

If healing is by first intention and the scar is supported by underlying sutures or is in a location where movement is slight, the skin sutures may safely be removed early. If healing is delayed for any reason, the skin sutures should be left in longer. In long, weight-bearing incisions (such as midline abdominal incisions), the sutures should remain in place at least 10 days and sometimes as long as 2 weeks.

When removing the sutures, one must take care to avoid contamination of the wound. This will be less of a problem if the knots are tied to one side of the incision. After all the sutures have been removed, the incision site should again be cleansed with an antiseptic.

postoperative management of pain

Analgesia

Analgesic agents are used to produce relief from pain without causing loss of consciousness. Although they are most commonly used postoperatively, they are also very useful in many non-operative situations. In order to appropriately determine when analgesics should be used, it is helpful to divide pain into three levels of magnitude: mild, moderate, and severe.

1.   **Mild pain** is a nuisance; it is tolerable and may or may not require analgesia. This level of pain does not interfere with the behavior of an animal and is difficult to identify. Treatment is not usually indicated. Pain categorized as mild may be associated with minor surgery, licking, or healing wounds. Mild pain can be produced manipulating the involved area of an animal; however, this area is not painful when at rest.

2.   **Moderate pain** starts to interfere with normal behavior, appetite, and activity. This level of pain is equated with that which would cause a human to seek relief, and would alter his/her activities such that they lose sleep, work less effectively, etc. In most cases, this level of pain in animals warrants relief. However, in an animal that has pain related to disc disease or a fractured bone, total abolition of pain may cause the animal to further injure itself. In many postoperative situations, pain can be minimized but not alleviated entirely. In the non-operative situation it is important to learn to recognize symptoms of moderate pain (decreased activity, decreased appetite, decreased tolerance to handling, loss of mobility, etc.) and examine the animal to determine if providing analgesia is appropriate.

3.   **Severe pain** might be defined as that which is intolerable. The animal may throw itself around the cage, pace, lose weight etc. In these cases, either administration of analgesics or euthanasia of the animal is indicated.

TABLE 14. DRUGS AND DOSES FOR ADMINISTERING
ANALGESIA TO HAMSTERS AND GERBILS

Drug[b]	Hamster	Gerbil
Butorphanol	1-5, q2-q4h	1-5, q2-4h
Buprenorphine	0.05-0.5, q8-12h	0.05-0.2,q8-12h
Pentazocine	10, q2-4h	10, q2-4h
Nalbuphine	4-8, q3h	4-8, q3h
Oxymorphone	0.2-0.5, q6-12h	0.3-0.5, q6-12h
Meperidine	20, q2-3h	20, q2-3h
Morphine	2-5, q2-4h	2-5, q2-4h
Flunixin	2.5, q12-24h	2.5,q12-24h

[b] All values are given in mg/kg IM or SC.[66]

postoperative care of animals

The surgeon or principle investigator is responsible for post-operative care of the animal. The investigator may be relieved of this responsibility when they have prearranged for the animals care to be tended for by another trained individual. If a surgery ends late at night and no previous transfer of responsibility has been make, the investigator remains responsible until this responsibility is passed on.

Effective postoperative care improves the animal's recovery. Postoperative care begins when the incision is closed and includes the following:

1. Treating all problems resulting from surgical manipulation, and maintaining nutritional balance and hydration.

2. Providing heat (hot-water bottle or circulating hot-water blanket) to maintain the animal's body temperature during recovery.

3. Monitoring the animal's vital signs (temperature and respiration) at regular intervals until the animal is awake and able to right itself.

4. Relieve pain when appropriate.

5. Monitor the animal daily for surgical wound healing and evidence of infection.

6. Evaluate the animal's temperature and respiration, appetite, fluid intake, feces and urine output, and character.

7. Dressing the wound (topical antibiotics, gauze to protect the wound, and bandaging) if needed.

---

euthanasia

Euthanasia is a term which by definition provides its own guidelines: **to produce a rapid and painless death.** Criteria that are essential for a painless death include rapid induction of unconsciousness followed by respiratory and/or cardiac arrest. For both humane and scientific reasons, these are guidelines that all investigators are obligated to uphold. Tables 15 and 16 present a list of clinical signs and findings that indicate the animal is in a morbid or moribund condition.

When to Euthanatize an Animal?

In most cases the research proposal defines the time when an animal must be euthanatized prior to study initiation. However, there are many instances when an animal becomes ill during a study and a decision must be made as to whether or not the animal is euthanatized or the medical problem is properly addressed. If medical treatment would interfere with the validity of the experiment or data interpretation, the animal should be euthanatized rather than permitted to suffer. The decision to allow the animal to continue to suffer with a medical problem without treatment is below acceptable standards of veterinary medicine. In addition, allowing an animal with a medical problem to remain on study will alter it's normal physiology, and will compromise the data being collected.

Guidelines for Recognizing Morbid and Moribund Animals

The decision as to when unscheduled euthanasia should be considered is difficult to make. Veterinarians are trained to recognize and diagnose normal health and well-being, as well as variations from normal in laboratory animals. When concerns arise regarding the health of an animal, veterinarians are available to assist investigators in determining the status and appropriate disposition of the animal. With their assistance, a timely and accurate decision regarding whether or not to treat or euthanatize an animal can be made.

The following lists, compiled from References 69, 70, 71, and 72, are provided as general guidelines regarding when an animal should be evaluated by a veterinarian for treatment or euthanasia. It is important to remember that no single sign is necessarily indicative of a serious disease. The entire clinical picture should be evaluated by a veterinarian in order to make the correct decision.

Signs of Morbidity (Disease or Illness) in Animals:

1. Slow, shallow, labored breathing.
2. Hunched posture.
3. Ruffled fur (rough hair coat).
4. Rapid weight loss (20 to 25% in 1 week).
5. Anorexia.
6. Hypothermia (often occurs with piloerection).
7. Hyperthermia.
8. Elevated respiratory rate.
9. Diarrhea or constipation.
10. Skin sores, infections, or necrotic tumors.
11. Decreased activity.
12. Change in behavior.

Selected Criteria for Euthanasia of Moribund (Dying) Animals:

1. Rapid weight loss (20 to 25% in 1 week).
2. Partial or sustained anorexia leading to extended weight loss and emaciation.
3. Impaired mobility restricting access to food and water.

4. Spreading hair loss caused by disease.

5. Rough coat, hunched posture, or distended abdomen leading to lethargy, especially if prolonged (3 days).

6. Diarrhea, especially if debilitating (3 days) or leading to emaciation.

7. Prolonged or intense diuresis leading to emaciation.

8. Persistent coughing, wheezing, and respiratory distress.

9. Distinct icterus (jaundice; yellow color to skin).

10. Persistent bleeding from any orifice.

11. Anemia leading to debilitation.

12. Rapid growth of a mass, or clinical signs of leukemia.

13. Paralysis.

14. CNS signs (head tilt, tremors, spasticity, seizures, circling, or weakness) especially if hindering the animal's ability to obtain food or water.

15. Persistent self-induced trauma.

16. Markedly discolored urine; increased or decreased urine volume.

17. Lesions interfering with eating, drinking, or ambulation.

18. Clinical signs of infection requiring a diagnostic necropsy.

19. Necrotic tissue or tumors.

20. Other clinical signs judged as being of moribund conditions.

Note: Regardless of the method of euthanasia selected, the person performing this procedure must subsequently ensure that the animal is dead. This may be done by opening the animals chest (thoracotomy) to create a pneumothorax. Preferably the thoracotomy is done on each side of the chest to ensure that all lung tissue deflates. This procedure must be done before the animal is disposed of.

common methods of euthanasia for hamsters and gerbils

A complete summary of recommendations for euthanasia can be found in the 1993 "Report of the American Veterinary Association Panel on Euthanasia.[73] Summarized below are common methods of euthanasia.

Barbiturate Overdose

Use of barbiturates requires that the facility have a Drug Enforcement Agency (DEA) registration for purchase and use of controlled substances. For euthanasia, pentobarbital is commonly administered, intraperitoneally, at a dose of 100 to 150 mg/kg. If a commercially available barbiturate euthanasia agent is used, the manufacturers directions should be followed.

Inhalant Anesthesia

Carbon dioxide is often used to euthanatize small rodents. Compressed gas is recommended over the use of dry ice. If dry ice is used, measures must be taken to prevent the animals from coming in contact with the dry ice.

Physical Methods

Cervical dislocation or decapitation may be used, provided they are approved by the IACUC and scientifically justified. Use of a plastic decapitation cone may be useful in restraining the animals while performing the procedure. A commercially available guillotine for rodents should be used. Before performing this procedure, you should consult with the veterinarian for specific training and instructions on this technique. Whenever possible, the animal should be sedated or anesthetized prior to performing cervical dislocation or decapitation.

notes

experimental methodology

Hamsters and gerbils have been used for a wide variety of experimental models. The purpose of this chapter is to describe the more common techniques used when handling hamsters and gerbils for experimental procedures. Hamsters and gerbils are not typically placed into mechanical restraint devices; therefore none of these devices are shown or described. The procedures shown in the chapter focus on hand restraint. As noted in earlier chapters, the personnel performing the experimental manipulations must be properly trained in the procedures.

hamster handling and restraint

Hamsters have a reputation for being aggressive. They generally exhibit such behavior after being startled, awakened, or handled roughly. Therefore, it is important to let the hamster know you are present before you handle them.

Methods for Picking a Hamster Up:

1. Hamsters may be picked up by the loose skin of the neck (Figure 37), but care must be taken to prevent them from turning and biting the handler.

2. Hamsters may be picked up by cupping them in your hands (Figure 38), or by gripping them over the back (similar to picking up a baseball).

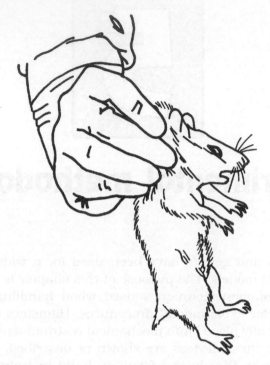

Fig. 37. Hamsters may be moved by lifting them by the loose skin over their back.

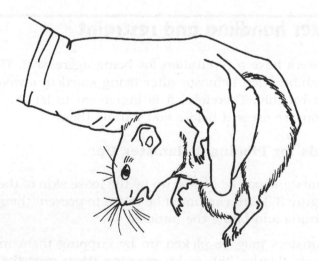

Fig. 38. Hamsters may be picked up by cupping them in your hand as shown.

3.   A protective glove may be worn for hamsters that are unaccustomed to handling; however, gloves may cause discomfort to the animals thereby developing an association between the glove, discomfort, and aggression.

Hamster Restraint

When hamsters are to be restrained, the loose skin of their neck must be gathered-up tightly to immobilize the animal (Figure 39). First, wake the hamster up if it is sleeping; a startled hamster is often aggressive. Next, remove the hamster from its cage and place it on a flat surface. Place your palm down over the hamster, with your thumb near the head. Slowly close your hand as you gather up the loose skin, being careful to avoid grasping the body. As you pick the hamster up you will find that it is securely held, and the skin is taut over the chest and abdomen.

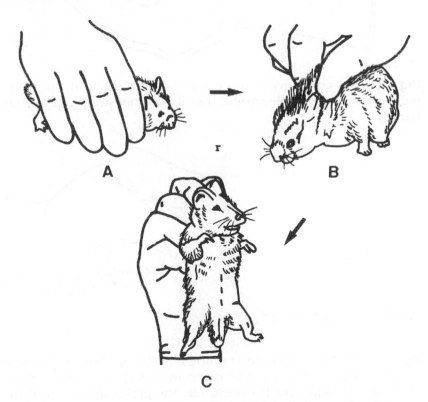

Fig. 39. Hamsters may be restrained by grasping the skin over their shoulders and pulling it tightly as they are picked up.

gerbil handling and restraint

Because the gerbil is gentile and docile, it is unnecessary to wear gloves to protect you from bite wounds when handling them. The main handling concern is that if startled by a loud or sudden noise, it may leap out of the cage and injure itself.

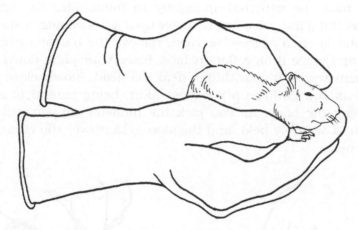

Fig. 40.   An easy way to move a gerbil is to scoop it up in the palm of your hands. Be careful that it does not jump out of your hands.

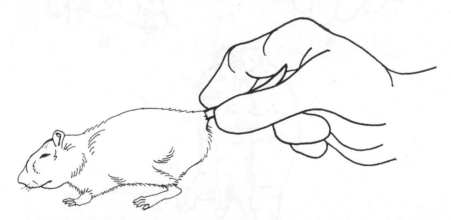

Fig. 41 and 42.   When restraining a gerbil, grab it by the base of its tail, then lift it up and reach over with your other hand to grasp the skin over it's neck. Lift the gerbil up and pull the skin over the neck tightly to restrain the animal.

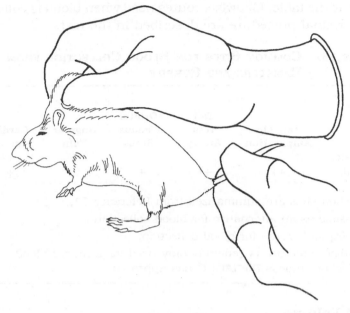

Fig. 42.

The primary method for handling a gerbil are to close your cupped hands over the animal and gently scoop it up in the palms of your hands (Figure 40), or **grasp the gerbil by the base of it's tail** (the tail skin may pull off if you grab the tail anywhere but at the base) with one hand and reach under the gerbil to lift and support it's body with your other hand. When restraining a gerbil, the tail base may be grabbed with one hand while reaching forward with the opposite hand to pick the gerbil up by the skin overlying the neck (Figures 41 and 42). Gerbils detest being placed on their back and will struggle if this is attempted.

sample collection

Blood Collection Sites

There is no preferred route for blood collection for the hamster or gerbil. The most common non-terminal blood collection routes used are summarized in Table 15, with a description of the procedures used for those and other routes included in the text following the table. Likewise, routes used when blood is collected as a terminal procedure are described in the text.

TABLE 15. COMMON SITES FOR BLOOD COLLECTION FROM HAMSTERS AND GERBILS

	Tail Tip Amputation	Tail Vein Artery	Orbital Venous Sinus	Jugular Vein	Cardiac Puncture
Hamster	–	–	+	+	++
Gerbil	++	++	++	+	++

Collection sites are summarized from Reference 74.

+ Possible as an alternative for blood collection.

++ Acceptable route for blood collection.

* Cardiac puncture is generally only used for terminal blood collection, unless the IACUC has approved

Blood Volume

Rapid removal of too much blood can produce hypovolemic shock. Removal of 10% of the animals blood volume will set off homeostatic responses (cholinergic), 15 to 20% will alter cardiac output and blood pressure, and 30 to 40% may induce hemorrhagic shock.[74] Repeated collections of small amounts of blood can result in the same effect. Therefore, if greater than 10% of the animal's circulating blood volume (1 ml/100 g body weight) is removed at a single collection point, replacement of the blood volume with a physiologic solution (e.g., Normosol R, 0.9% saline, lactated Ringer's solution) warmed to body temperature is recommended. If repeated blood samples are collected from the same animal over the course of 1 to 2 days or on a weekly basis, the amount of blood per sample should be decreased to a maximum of 1.0% of the animals circulating volume every 24 hours. This will permit the animal to begin regenerating blood and recover from the blood loss. Once again, donor blood or physiologic fluid (e.g., Normosol R, 0.9% saline, lactated Ringer's solution) can be given to minimize the stress of blood depletion and hypovolemia. The animal's hematocrit can also be monitored to be sure it does not become anemic.

Although the hamster and gerbil are reported to have a blood volume equivalent to 7.8% and 6.7% of their body weight,[74] respectively, for this discussion we will assume that total blood volume of both is equal to approximately 6% of the animal's body weight. (For example, a 100-gram gerbil has a total blood volume of 6%, which equals 0.06 × 100 g = 6 g blood = 6 ml total blood volume.) By using this assumption, a margin of safety is built into calculations for the maximum amount of blood what can be withdrawn safely.

When collecting all of the blood possible from an animal ("exsanguinating or bleeding out"), you can expect to collect 1/2 of the animal's total blood volume. For example, if a 100 g gerbil has a total blood volume of 6 ml: 1/2 or 3 ml (3% of body weight) is available when exsanguinated.

Note: Animals should only be exsanguinated while under general anesthesia.

Blood Collection Vials

When collecting blood, a variety of blood collection vials can be used (see Figure 43). When plasma or cellular constituents are desired, use of tubes containing an appropriate anticoagulant such as EDTA is recommended (lavender-top tube). Tubes containing heparin as an anticoagulant may also be useful. If blood glucose is to be measured, sodium fluoride is often added to the collection tube.

When serum is required, a red-top tube is used. This tube will separate the clot from the serum.

Blood Collection

hamsters

Possible routes for non-terminal blood collection are listed above. The procedures used for these routes are similar to those described for rats and mice.

1. Collection of blood from the **orbital sinus** should be performed only with the animal anesthetized. Once the animal is anesthetized, it is held firmly by the scruff of the neck on a solid surface, allowing the eye to protrude. It is important not to obstruct the animal's ability to breathe when performing this procedure. A heparinized

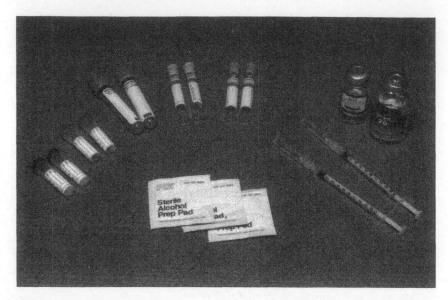

Fig. 43. A variety of blood collection vials are used as shown above. Anesthesia may be required for the blood collection. The blood collection site should be decontaminated with a prep pad.

    hematocrit tube or a Pasteur pipette is inserted into the lateral canthus of the eye and directed medially and posterior, toward the back of the orbit, until it hits the bony wall of the eye socket and penetrates the venous sinus. Blood is then collected from the free tip of the capillary tube. Place gentle pressure on the eye following collection to stop the bleeding. Application of a topical ophthalmic anesthetic such as proparacaine to the eye will provide relief from post-procedural pain, and is highly recommended.

2.   The **lateral tarsal vein**, which lies on the lateral side of the hind foot, may be used for blood collection. A 25 gauge or smaller needle is inserted into the occluded vein. As blood flows into the hub, it can be collected in a hematocrit tube or a blood collection tube.

3.   The **abdominal vena cava or abdominal aorta** may be used in anesthetized animals as a terminal blood collection procedure. This vein lies directly below the spinal column within the abdominal cavity. A ventral midline

incision of the abdomen, with reflection of the intestinal tract to one side or the other, will expose this vein.

4. **Because of the potential for causing post-procedural problems such as cardiac tamponade or a collapsed lung, cardiac puncture is generally only used for terminal blood collection, unless the IACUC has approved this method as a recovery procedure. When cardiac puncture is used anesthesia is always required.** The procedure may be done with the animal laying on its side or back. With the animal on its back, the needle is advanced to the left of the xiphisternal junction at a 30 degree angle down into the heart (Figure 44). The needle should be directed at the apex beat. The heart can also be approached from the right or left side at the 4th to 5th intercostal junction, with the needle being directed at the apex beat.

5. **Tail tip clipping** is not used in the hamster. However, **toenail clipping** may be used to collect small amounts of blood. Pressure must be applied to the site of collection until bleeding stops.

gerbils

1. Blood can be collected from unanesthetized animals from the **lateral tail vein.** This is done by placing pressure on the tail vein by gently squeezing the base of the tail to restrict the flow of blood out of the vien. A needle is then inserted into the vein. It may be necessary to dip the gerbil's tail in warm water (11°F) for 10 to 20 seconds to promote blood flow. In addition, placing the animal in a thermostatically warmed chamber (30 to 35°C for 10 to 15 minutes) will result in vasodilation.

2. Collection of blood from the **retro-orbital sinus** from the medial canthus can be used. As with the hamster, this method should include some form of short-term sedation or anesthesia, and the procedure is carried out as described with the hamster. Application of a topical ophthalmic anesthetic should also be considered to provide post-procedural analgesia.

3. **Cardiac puncture** may also be used as described with the hamster and should be reserved for terminal bleeding unless approved otherwise by the IACUC. Similarly, collection of blood from the abdominal vena cava or abdominal aorta may be used and is a terminal procedure that should only be done under anesthesia.

> **Note:** Cardiac puncture should only be used if approved by the animal care and use committee. When cardiac puncture is performed, anesthesia is always required.

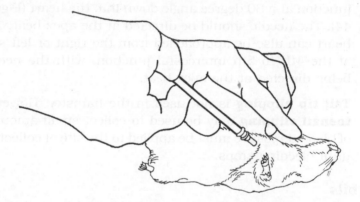

Fig. 44. The cardiac puncture procedure may be done with the animal on its side or back. With the animal on its back, the needle is advanced to the left of the xiphisternal junction, downward at a 30 degree angle into the heart. The plunger on the syringe should be withdrawn slightly so that blood will be pulled into the syringe when the heart is entered.

4. **Tail-tip clipping or toenail clipping** may be used to collect small amounts (0.1 to 0.2 ml) of blood. The tail tip should only be used 1 or 2 times, otherwise the coccygeal vertebrae will become involved. Pressure must be applied to the site of collection until bleeding stops.

5. The gerbil may be exsanguinated from the **abdominal vena cava or abdominal aorta** as described for the hamster.

Urine and Feces Collection

Usually urine is collected in directly from the bladder at necropsy, or using a metabolism cage. When using a metabolism cage, note that since hamsters and gerbils void a small volume of urine, it may evaporate during the collection period, thus affecting the specific gravity and possibly other parameters.

Feces are usually collected as a fresh sample from the metabolism cage pan, or directly from the colon at necropsy.

test article/compound administration

For both the hamster and gerbil, with the exception of intravenous injections, most injections and intubations are done as in the rat or mouse. The skin over the injection site is often cleaned with an alcohol wipe before administering the injection. Compounds can also be given by parenteral or oral routes.

Parenteral Routes

intravenous injections

Needles of 23 to 25 gauge are typically used for most injections. However, for very viscous substances, a larger needle (21 gauge) may be required. Prior to administering any injection, once the needle is inserted into the area the injection is to be given, it is appropriate to draw back on the syringe plunger to determine if the needle is inserted into a blood vessel. If, upon drawing back the plunger, you see a flash of red in the needle hub, you are in a vessel and may proceed to give the injection.

Since **hamsters** lack an easily accessible tail vein, most IV injections are made in the cephalic vein (anterior forearm), tarsal vein (lateral side of hind limb), lingual vein, orbital venous sinus or jugular vein (requires anesthesia and a surgical cutdown). In addition, since repeated IV injections are so difficult in the hamster, intracardiac injections may be used, provided the procedure is approved by the IACUC.

When using the cephalic vein, which runs along the anterior medial aspect of the forearm the animal is anesthetized and a rubber band tourniquet is applied at the elbow, causing the vein to enlarge and engorge. Once the vein is engorged, the needle is inserted at a 10 degree angle, and an injection of up to 0.3 ml is easily given. No reinjections are given for 24 hours.[75]

With the jugular vein, the skin overlying the vein must be prepared for a surgical cutdown. A veterinarian should be consulted before this procedure is used. Likewise, a veterinarian should be consulted if the lingual vein or orbital venous sinus are used for injection, and these procedures should only be performed on anesthetized animals.

Intracardiac injections are generally given with the animal laying anesthetized on its back. The needle is inserted as described for cardiac bleeding, and the injection given slowly. With proper training, repeated intracardiac injections are possible.

In **gerbils**, the lateral tail vein is accessible and commonly used for IV injections. It may be necessary to dip the gerbil's tail in warm water (11°F) for 10 to 20 seconds to promote vasodilatation for easier injections, or use the heating chamber as described previously for blood collection. Once the vein becomes dilated, the base of the tail is firmly gripped and a needle (23 gauge or smaller) is inserted into the vein. When a flash of blood is seen in the hub of the needle, the injection is given. If the injection was given correctly, the tail vein will temporarily blanch

subcutaneous injections

Most subcutaneous injections are easily given between the shoulder blades in both hamsters and gerbils.(Figure 45). The skin is tented with one hand and the needle is inserted under the skin. Once this is done, draw back on the syringe plunger to determine if you have inserted the needle into a bleed vessel. If you see a flash of red in the needle hub, you are in a vessel and must withdraw the needle and place it in a new location. If you do not see blood when drawing back on the syringe plunger, the injection may be given. Since hamsters are restrained by the skin over the shoulder blade, subcutaneous injection may also be given in the inguinal region.

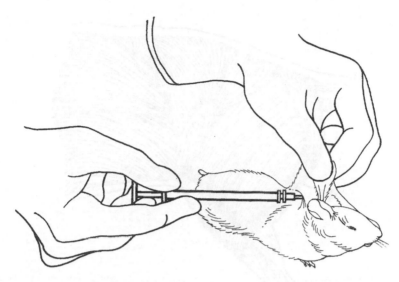

Fig. 45. Most subcutaneous injections are given between the shoulder blades. The skin is tented with one hand and the needle is inserted under the skin.

intramuscular injections

The caudal thigh muscles are most frequently used for intramuscular injections as depicted in Figures 46 and 47. Care must be taken to avoid injecting into the sciatic nerve that runs laterally along hip and the hind limb.

intraperitoneal injections

Intraperitoneal injections are generally given in the lower left or right abdominal quadrant while the animal's head is tilted downward (Figure 48). Once the syringe is placed in the abdominal cavity, the plunger is slightly withdrawn. If any intestinal contents are seen in the syringe (brown or green liquid, or urine), the needle should be withdrawn and administration attempted again with a new needle and syringe. Since the gerbil detests being placed on it's back, this procedure must be modified to prevent them from struggling. The gerbil is restrained as previously described, and while held vertically, the needle is slipped into the abdominal quadrant.[13]

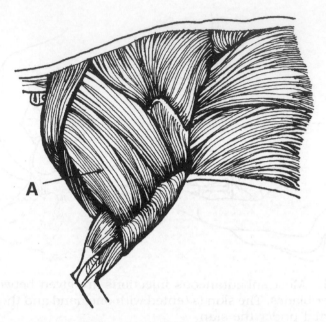

Fig. 46.   When giving an intramuscular injection, the injection is given in the caudal thigh muscles (depicted with the overlying tissue removed) in an area (A) that will avoid injecting into the sciatic nerve.

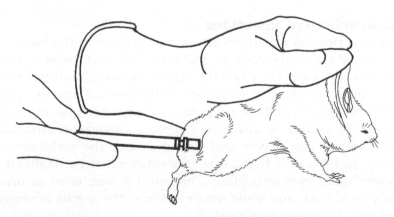

Fig. 47.   This figure shows the restraint method and injection site for administering an intramuscular injection.

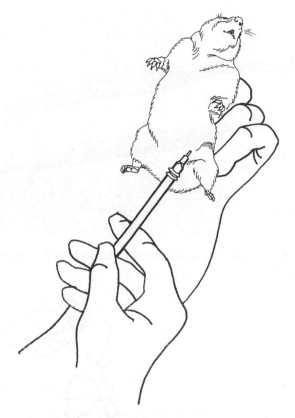

Fig. 48. An intraperitoneal injection is typically given in the lower left or lower right quadrant of the abdomen. It is important to withdraw the plunger before making the injection. Any material withdrawn into the syringe would indicate that the needle has entered an abdominal structure and must be withdrawn, changed and redirected before the injection is given.

Oral Route

gavage

Dosing hamsters and gerbils with a gavage needle (stainless steel stomach tube with a ball tip: Figure 49) is performed while the hamster or gerbil is restrained as previously described. A 18- or 20-gauge, stainless-steel ball-tipped gavage needle of approximately 10-12 cm long is used. Before inserting the needle, the distance between the mouth and stomach (last rib) is measured (Figure 50). This gives you a general idea of how far the tube must be inserted to reach the stomach. Once the animal

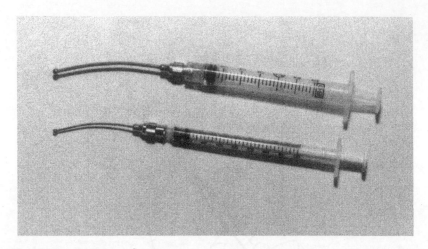

Fig. 49. Ball-tipped gavage needles attached to syringes used for oral dosing.

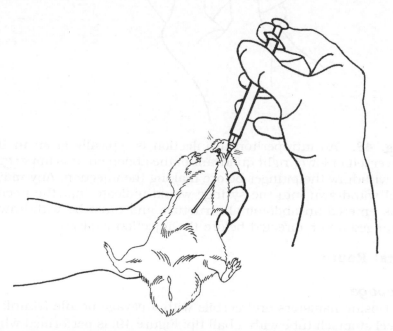

Fig. 50. The distance from the oral cavity to the stomach is measured prior to dosing the animal. If resistance is met when inserting the needle, do not advance the needle because you have most likely entered the trachea; withdraw the needle and begin again.

is restrained, the needle is slid along the hard palate and directed posteriorly through the esophagus and into the stomach (Figure 51). When the needle has been inserted to the previously measured depth, the dose can be given.

Note: If resistance is met when inserting the needle, do not advance the needle any further because you are most likely entering the trachea. Withdraw the needle and begin again.

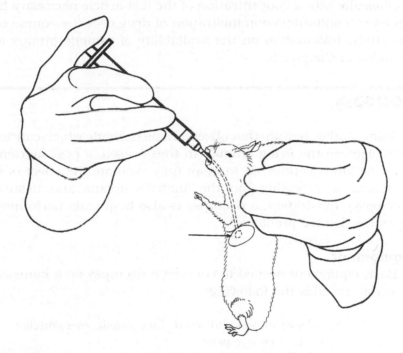

Fig 51. The needle should be advanced all the way into the stomach before the dose is adminstered.

hamster cheek pouch

Fluids in small amounts (0.1 ml), and even tumors and tumor cells, may be instilled into the hamster's cheek pouch. In order to implant tumors, the pouch must be everted and the injection is given. This process becomes remarkably easier when the hamster is anesthetized (see Figure 4 in Chapter 1).

Water and Food

Test articles may be incorporated into the water and/or food supply as long as they do not significantly alter palatability and intake.

Implantable Osmotic Pumps

Implantable osmotic pumps are becoming more popular as a means to administer compounds for periods up to 28 days. These pumps are surgically implanted subcutaneously or intra-peritoneally with a concentration of the test article necessary to achieve a continuous administration of drug over the course of the study. Information on the availability of osmotic pumps is included in Chapter 6.

necropsy

Many studies require that observations be made which cannot take place on the living animal. In these cases, a post mortem examination is conducted to more fully evaluate the effect of a treatment or procedure on the animal's organs and tissues (necropsy). In addition, a necropsy is also frequently performed to assess disease problems.

Equipment

Basic equipment needed to conduct a necropsy on a hamster or gerbil includes the following:

1. Latex or rubber gloves, lab coat, face mask, eye goggles, or other protective eyewear.

2. A small metric ruler.

3. Toothed and serrated tissue forceps, and a probe.

4. Scalpel blades and handles.

5. Dissecting and small operating scissors.

6. A dissecting board (plastic or cork).

7. Bone-cutting forceps.

8. Sterile swabs for bacteriological culture of tissues.

9.   Large (18-gauge) and small bore (25-gauge) hypodermic needles and syringes (1, 3, and 5 ml).

10.   Saline for washing of structures, and paper towels for absorption of blood and other fluids are useful. Additional equipment may be useful and can be added to this basic kit.

11.   A method to sterilize instruments (open flame or glass bead sterilizer)

The postmortem evaluation necropsy should be performed in a dedicated necropsy room and on a surface that will facilitate drainage of blood and fluids, and that can be easily cleaned and sanitized. Since necropsy is often used to determine if an animal from the colony is harboring an infectious agent which may pose a threat to the health of the remaining animals in the facility, the necropsy room should be restricted to entry of personnel that do not have to reenter the animal facilities on the day of the necropsy. If they do have to reenter the animal facility, those individuals should wear protective clothing in the necropsy room and change this clothing when leaving the room.

Stainless-steel necropsy tables with a down-draft airflow are ideal, since they will exhaust infectious agents and noxious odors away from personnel. These tables should be certified annually to ensure that they are working adequately. If this type of equipment is unavailable, an area that is isolated from other animals, personnel areas, the surgery suite and feed and bedding storage could be used, provided that the area can be appropriately cleaned and sanitized following each use.

A 10% neutral buffered formaldehyde mixture is the most commonly used tissue fixative. When using this fixative, recognize that it can cause allergic reactions and irritation of surfaces lined by mucous membranes,[76,77] and is considered to be a human carcinogen.[78] Thus, exposure of personnel to formaldehyde must be limited by providing adequate ventilation of the necropsy and tissue processing areas.

Personnel conducting necropsies should wear a clean lab coat, latex or rubber gloves, protective booties and a face mask, and protective eyewear. Although specific pathogen-free hamster and gerbils harbor few zoonotic infectious agents, this equipment will further decrease exposure of personnel to airborne allergens and formaldehyde, as well as protect clothes from soiling with blood or other material. Whenever possible, the clothing noted above should be disposable.

necropsy technique

Ideally, the hamster and gerbil should be necropsied immediately after death. Alternatively, carcasses may be stored for a short time (several hours) under refrigeration to delay tissue decomposition. Carcasses should not be stored in refrigerators used for storage of food for animals or personnel. Freezing of carcasses can significantly interfere with meaningful necropsy. An in-depth description of anatomy and some details on necropsy methods for the hamster and gerbil can be found elsewhere.[79,80] General procedures for necropsy of a hamster or gerbil are as follows:

1. The animal's body weight is obtained and recorded, and is examined externally for abnormalities such as discoloration, hair loss, wounds, masses, discharges (nasal or ocular), and fecal or urine staining. In addition, the oral cavity is examined, with particular attention paid to the teeth. Any cultures of the external surface lesions or discharges should be taken.

2. The animal is secured to a necropsy board. The skin is incised along the ventral midline with the scalpel blade or scissors, beginning at the lower jaw and continuing along the midline caudally to the pubis. Using the scalpel, or scissors, the skin is reflected laterally away from the subcutaneous tissues and underlying musculature. These tissues are examined.

3. The abdominal wall is lifted up with forceps and incised. Using scissors, the abdominal cavity is opened along the ventral midline.

4. The abdominal organs and the peritoneal surfaces are examined for abnormal hydration, coloration, size, and presence of masses, traumatic damage, or any other abnormal appearance. If fluid is present in an excessive

amount or with an abnormal color, it should be sampled for cytology and bacterial culture, and the volume and appearance of such fluids noted and recorded. Depending on the time between death and necropsy, and the carcass storage conditions, tissues may appear abnormal due to postmortem autolysis, a non-disease process in which tissues degrade after the animal has died.

5. The thoracic cavity is exposed by reflecting the liver forward and cutting the diaphragm. Once this is done, clip the ribs using the bone cutting forceps. The clipped portion of the rib cage is then lifted off and removed or reflected laterally.

6. The lungs, heart, and pleural surfaces are examined for abnormalities. The organs are removed for inspection by cutting the trachea and cutting all attachments of trachea, lungs, and heart caudally to the diaphragm.

7. Abnormal fluids found in the abdominal or thoracic cavity should be sampled for cytology and bacterial culture, and the volume and appearance of such fluids noted and recorded.

8. Other masses or abnormal tissues can be cultured using bacteriological culture swabs if infection is suspected.

9. Samples of tissues can be preserved in 10% neutral buffered formalin and saved for later processing and evaluation. Smaller samples (a few centimeters thick) allow quicker penetration of formalin, better preservation, and are therefore preferred.

notes

resources

This chapter provides a list of sources for obtaining information on hamsters and gerbils, as well as lists of resources for obtaining information on equipment and materials, examples of vendors, and organizations dealing with hamsters and gerbils. The lists are not exhaustive, and they do not imply endorsement of one vendor over another. What they do provide is a starting point for developing a list of resources. Sources for equipment are provided below with contact information provided at the end.

organizations

Several professional organizations serve as a primary contact for obtaining information on distinct issues related to the care and use of laboratory hamsters and gerbils. Membership in these organizations should be considered, since it allows the laboratory animal science professional to remain knowledgeable of regulatory issues, improved procedures for the use of animals, management issues, and animal health issues. Several of the relevant organizations are listed:

The **American Association for Laboratory Animal Science (AALAS)**, 70 Timber Creek Drive, Cordova, TN 38018 (Tel: 901-754-8620). AALAS serves all members of the laboratory animal science community, including but not limited to principal investigators, animal care technicians, and veterinarians. The journals, *Laboratory Animal Science* and *Contemporary Topics in Laboratory Animal Science*, and the technicians newsletter *Tech Talk*, are published by AALAS. AALAS sponsors a program for certification of laboratory animal science professionals at 3 levels: Assistant Laboratory Animal Technician (ALAT), Laboratory Animal Technician (LAT), and Laboratory Animal Technologist (LATG). The AALAS-affiliated Institute for Laboratory Animal Management (ILAM) is a program designed to provide state of the art training in laboratory animal facility management. In addition, the association sponsors an annual meeting and an electronic bulletin board (COMPMED). Local groups have also organized into smaller branches.

The **Laboratory Animal Management Association (LAMA)** provides an opportunity for information exchange between individuals who's primary responsibility is managing laboratory animal facilities. LAMA promotes the dissemination of ideas, knowledge and experiences, encourages continued education, and assists in training managers of animal facilities. The association publishes the *The LAMA Review* on a quarterly basis and sponsors periodic meetings. The contact for LAMA changes annually with the elected president. The current contact for LAMA may be obtained from AALAS.

The **American Society of Laboratory Animal Practitioners (ASLAP)** is an association of veterinarians engaged in laboratory animal medicine. The society publishes a newsletter and sponsors annual meetings, generally in conjunction with annual meetings of AALAS and the American Veterinary Medical Association (AVMA). The contact for ASLAP changes annually with the elected president. Current contact information may be obtained from AALAS.

The **American College of Laboratory Animal Medicine (ACLAM)** is an association of laboratory animal veterinarians founded to encourage education, training, and research in laboratory animal medicine. ACLAM is recognized as a sub-specialty of veterinary medicine by the AVMA. The ACLAM Board of Directors annually certifies veterinarians as Diplomates in laboratory animal medicine by means of an examination, experience, and publication requirements. The group sponsors the ACLAM Forum as well as sessions at the annual AALAS meeting. Contact is established through AALAS or the AVMA.

The **International Council for Laboratory Animal Science (ICLAS)** promotes and coordinates the development of laboratory animal science throughout the world. ICLAS sponsors international meetings every fourth year, with regional meetings being held on a more frequent basis. The organization is composed of national, scientific, and union members.

The **Institute of Laboratory Animal Resources (ILAR)**, under the auspices of the National Research Council, develops and makes scientific and technical information available on laboratory animal models and other biologic resources. Useful publications available from ILAR include the *Guide for the Care and Use of Laboratory Animals* and the *ILAR Journal*. Contact with ILAR can be established at 2101 Constitution Avenue, NW, Washington, DC 20418 (Tel: 202-334-2590).

The **Association for Assessment and Accreditation of Laboratory Animal Care International, Inc. (AAALAC International)** is a nonprofit organization that provides peer evaluation of laboratory animal care and use programs. AAALAC accreditation is widely accepted as evidence of a high-quality animal care and use program. Contact with AAALAC may be made through the Executive Director at 11300 Rockville Pike, Suite 1211, Rockville, MD 20852-3035 (Tel: 301-231-5353).

publications

A number of published materials are valuable as additional reference materials, including both books and periodicals.

Books
The following books may be worthwhile sources of additional information:

1. ***A Colour Atlas of Anatomy of Small Laboratory Animals: Rat, Mouse and Hamster,*** by P. Popesko, V. Rajtoua, and J. Horak, 1990. Wolfe Publishing, Ltd. London, England.

2. ***The Anatomy of the Mongolian Gerbil (Meriones unguiculatus)***, by W. M. Williams, 1974. Tumblebrook Farms, Inc., Westbrookfield, MA, 01585.

3. ***The Biology and Medicine of Rabbits and Rodents,*** by J. E. Harkness and J. E. Wagner, 1995, Williams & Wilkins, Baltimore, MD 21298-9724.

4. ***Formulary for Laboratory Animals***, by C.T. Hawk and S. L. Leary, 1995. Iowa State University Press, Ames, IA 50014.

5. ***Laboratory Animal Anesthesia***, by P.A. Flecknell, 1987. Academic Press, Inc., 525 B Street, Suite 1900, San Diego, CA 92101.

6. ***Handbook of Veterinary Anesthesia***, by W. W. Muir, J. A. E. Hubbell, R. T. Skarda, and R. M. Bednarski 1995. C.V. Mosby Co., 11830 Westline Industrial Drive, St. Louis, MO 63146.

7. ***Necropsy Guide: Rodents and the Rabbit,*** by D.B. Feldman and J.C. Seely, 1988. CRC Press, Inc., 2000 Corporate Blvd. N.W., Boca Raton, FL 33431.

8. ***Laboratory Hamsters***, by Van Hoosier, G. L. and McPherson, C. W., Academic Press, Orlando, FL 1987.

Periodicals

The following periodicals are excellent sources of current relevant information:

1. ***Laboratory Animal Science.*** Published by the American Association for Laboratory Animal Science. For contact information, see the above listing for AALAS.

2. ***Contemporary Topics in Laboratory Animal Science.*** Published by the American Association for Laboratory

Animal Science. For contact information, see the above listing for AALAS.

3. ***Laboratory Animals***. Published by Royal Society of Medicine Press, 1 Wimpole Street, London W1M 8AE, UK.

4. ***Lab Animal.*** Published by Nature Publishing Co., 345 Park Avenue South, NY 10010-1707.

5. ***ILAR Journal.*** Published by the Institute of Laboratory Animal Resources, National Research Council. For contact information, see the above listing for ILAR.

6. ***The Gerbil Digest***, (Figure 52) published by Tumblebrook Farms, Inc., West Brookfield, MA 01585. This periodical is out-of-print, but old issues can be found at some libraries.

Fig. 52. *The Gerbil Digest,* published by Tumblebrook Farms, Inc., contains many details on gerbils used in research. (copies of *The Gerbil Digest* were kindly provided by D. Robinson.)

electronic resources

Many online sources of information relevant to the care and use of laboratory animals, including hamster and gerbils, are available. These include:

1. **Comparative Medicine Discussion List (COMPMED).** An electronic mailing list available through the Internet, COMPMED is a valuable means to quickly tap into the expertise of laboratory animal science professionals around the world. In addition, COMPMED archives all correspondence, and the archives may be searched for reference information. At the time of publication, those individuals who are interested in using this resource should subscribe to **listserv@listserv.aalas.org** and mail to **compmed@listserv.aalas.org**.

2. **Network of Animal Health (NOAH).** NOAH is a commercial online service sponsored by the American Veterinary Medical Association. Forums cover a variety of topics, some of which would be of interest to those caring for and using hamsters and gerbils. Additional information can be obtained from the American Veterinary Medical Association (1931 N. Meacham Rd., Suite 100, Schaumburg, IL; 1-800-248-2862; e-mail: 72662.3435@compuserve.com).

purchasing hamsters and gerbils

Hamster and gerbils may be obtained from vendors of varying size and quality. Since hamsters and gerbils may contract a variety of diseases that could alter the validity of research results, the purchase of only specific pathogen-free (SPF) hamsters and gerbils is strongly encouraged. Vendors should be asked to supply recently obtained information regarding the health status of their hamster and gerbil colonies prior to your purchase.

Small local or regional vendors frequently offer quality animals at reasonable prices, however, they may not have the same professional resources or quality control behind them that a larger supplier has. In addition, large vendors are often a good source of high-quality hamster and gerbils with known health status. Since it is not possible to list all of the hamster and gerbil vendors here, the following are examples of vendors that supply hamsters and gerbils:

1. Bio Breeders Inc., 280 Sheldon Road, Fitchburg, MA 01420-1818 USA, (Tel: 508-343-3000) **for hamsters.**

2. Charles River Laboratories, Inc., 251 Ballardvale Street, Wilmington, MA 01887 USA (Tel: 800-522-7287) **for hamsters and gerbils**.

3. Harlan Sprague Dawley, Inc., Box 29176, Indianapolis, IN 46229-0176 USA, (Tel: 317-894-7521) **for hamsters and gerbils**.

4. Simonsen Laboratories, Inc., 1180-C Day Road, Gilroy, CA 95020-9308 USA, (Tel: 408-847-2002) **for hamsters**.

feed

Several large feed vendors such as those listed below provide a high quality, nutritionally balanced diet. However, just as with animals, analysis of the feed for contamination during production, evaluation of the storage conditions at the supplier or a regional distributor's warehouse, and assessment of conditions during delivery is critical to ensure the provision of high-quality feed.

1. Bio-Serv, Inc., PO Box 450, 8th & Harrison Streets, Frenchtown, NJ 08825 (Tel: 1-800-473-2155).

2. Harlan Teklad, Inc., P.O. Box 44220, Madison, WI 537444220 (Tel: 1-800-483-5523).

3. PMI/Purina Mills, Inc., 505 North 4th St., PO Box 548, Richmond, IN 47375 (Tel: 1-800-227-8941).

4. United States Biochemical Corp., PO Box 22400, Cleveland, OH 44122 (Tel: 1-800-321-9322).

equipment

Sanitation

Several sources of disinfectants and other sanitation supplies are listed below.

1. BioSentry, Inc., 1481 Rock Mountain Blvd., Stone Mountain, GA 30083-9986 (Tel: 1-800-788-4246).

2. Pharmacal Research Labs, 33 Greater Hill Road, Naugatuck, CT 06770–369 (Tel: 1-800-243-5350).

3. Quip Laboratories Ind., 1500 Eastlawn Avenue, Wilmington, DE 19802 (Tel. 302-761-2600).

4. Rochester Midland, Inc., 333 Hollenbeck St., P.O. Box 1515, Rochester, NY 14603-1515 (Tel: 1-800-836-1627).

5. Steris Corporation, Steris Scientific, 5960 Heisley W. Road, Mentor, OH 44060 (Tel: 1-800-444-9009).

Cages and Research and Veterinary Supplies

Sources for pharmaceuticals, hypodermic needles, syringes, surgical equipment, bandages, and other related items are provided below. Unless you have extensive experience with the pharmaceuticals you plan to order, they should be ordered and used only under the direction of a licensed veterinarian. Cages should meet the size requirements as specified by relevant regulatory agencies. Stainless steel is preferable to galvanized steel.

TABLE 11. POSSIBLE SOURCES OF EQUIPMENT AND SUPPLIES

Item	Source
Cages and supplies	1, 2, 4, 6, 11, 12, 13, 14
Veterinary and surgical supplies	5, 7, 8, 9, 10, 16, 17
Gas anesthesia equipment	9, 15, 16, 17
Syringes and needles	5, 7, 8, 10, 17
Osmotic pumps	3
Necropsy tools	5, 8, 16

contact information for cages and research and veterinary supplies

1. Allentown Caging Equipment, Inc., PO Box 698, Allentown, NJ 08501-0698 (Tel: 609-259-7951 or 1-800-762-2243).

2. Alternative Design Manufacturing and Supply, Inc., 16396 Highway 412, Siloam Springs, AR 72761 (Tel: 1-800-320-2459).

3. Alza Corporation, 950 Page Mill Road, PO Box 10950, Palo Alto, CA 94303-0802 (Tel: 1-800-692-2990).

4. Ancare Corp., 2475 Charles Court, PO Box 661, North Bellmore, NY 11710 (Tel: 1-800-645-6379).

5. VWR Scientific Products Corporation, 1310 Goshen Parkway, West Chester, PA 19380, (Tel: 1-800-932-5000).

6. Britz-Heidbrink, Inc., PO Box 1179, Wheatland, WY 82201-1179 (Tel: 307-322-4040.

7. Butler Co., Inc., 5000 Bradenton Ave., Dublin, OH 43017 (Tel: 1-800-225-7911).

8. Fisher Scientific, Inc., 711 Forbes Ave., Pittsburgh, PA 15219-4785 (Tel: 1-800-766-7000). 12. Harvard Apparatus, 22 Pleasant St., South Natick, MA 01760 (Tel: 1-800-272-2775).

9. Harvard Apparatus, 22 Pleasant Street, South Natick, MA 01760 (Tel: 1-800-272-2775).

10. IDE Interstate, Inc., 1500 New Horizons Blvd., Amityville, NY 11701 (Tel: 1-800-666-8100).

11. Lab Products, Inc., 255 West Spring Valley Ave., PO Box 808, Maywood, NJ 07607 (Tel: 201-843-4600 or 1-800-526-0469).

12. Lenderking Caging Products, Inc., 1000 South Linwood Ave., Baltimore, MD 21224 (Tel: 410-276-2237).

13. Lock Solutions, Inc., P.O. Box 611, Kenilworth, NJ 07033 (Tel: 1-800-947-0304).

14. Otto Environmental, 6914 N. 124th St., Milwaukee, WI 53224 (Tel.: 1-800-484-5363 Ext. 6886).

15. Vetamac, Inc., PO Box 178, Rossville, IN 46065 (Tel: 1-800334-1583).

16. Viking Products, Inc., PO Box 2142, Medford Lakes, NJ 08055 (Tel: 609-953-0138).

17. J. A. Webster, Inc., 86 Leominster Road, Sterling, MA 01564 (Tel: 1-800-225-7911).

bibliography

1. Hobbs, K. R., Hamsters, in *The UFAW Handbook on The Care & Management of Laboratory Animals*, 6th edition, Poole, T., Longman Scientific and Technical, Essex, U.K., 1987, chap. 23.

2. Clark, J. D., Historical perspectives and taxonomy, in *Laboratory Hamsters*, Van Hoosier, G. L. and McPherson, C. W., Academic Press, Orlando, FL, 1987, chap. 1.

3. Magalhaes, H., Foreword, in *Laboratory Hamsters*, Van Hoosier, G. L. and McPherson, C. W., Academic Press, Orlando, FL, 1987, p xii.

4. Jones, C. H. and Pinel, J. P., Linguistic analogies and behavior: the finite-state behavioral grammar of food hoarding in hamsters, *Behav. Brain Res.*, 36, 189-97, 1990.

5. Bivin, W. S., Olsen, G. A., and Murray, K. A., Morphophysiology, in *Laboratory Hamsters*, Van Hoosier, G. L. and McPherson, C. W., Academic Press, Orlando, FL, 1987, pp. 10-36.

6. Maghalaes, H., Gross anatomy, in *The Golden hamster: Its Biology and Use in Medical Research*, Hoffman, R. A., Robinson, P. F. and Magalahaes, H., Iowa State University Press, Ames, IA, 1968, pp. 91-109.

7. Wagner, J. E. and Farrar, P. L., Husbandry and medicine of small rodents in *Vet. Clin. N. Amer., Sm. Anim. Prac.*, 17, 1061-1087, 1987.

8. Robinson, D. G., Gerbile ecology of the Mongolian gerbil, *The Gerbil Digest*, 3, No. 2, 1-2, 1976.

9. Schwentker, V., The gerbil: a new laboratory animal, *The Illinois Veterinarian*, 6, 5-9, 1963.

10. Norris, M. L., The gerbil, in *The UFAW Handbook on The Care & Management of Laboratory Animals*, 6th edition, Poole, T., Longman Scientific and Technical, Essex, U.K., 1987, Chap. 22.

11. Robinson, D. G., Gerbil classification and nomenclature. *The Gerbil Digest*, 2, No.1, 1-4, 1975.

12. Charles River Laboratories, Kingston, NY, Personal communication, 1996.

13. Rollin, B. E. and Kesel, M. L., *The Experimental Animal in Biomedical Research. Volume II: Care, Husbandry, and Well being. An Overview by Species.* CRC Press, Boca Raton, FL, 1995.

14. Turner, J. W. and Carbonell, C. A relationship between frequency of display of territorial marking behavior and coat color in the Mongolian gerbil, *Lab. Anim. Sci.*, 34, 488-490, 1984.

15. Holmes, D. D., *Clinical Laboratory Animal Medicine: an Introduction*, The Iowa State University Press, Ames IA, 1984.

16. Robinson, D. G., The Mongolian gerbil: anatomical studies, *The Gerbil Digest*, 3, No. 3, 1-4, 1976.

17. Vincent, A. L., Rodfick, G. E., and Sodeman, W. A., The pathology of the Mongolian gerbil (*Meriones ungiculatus*: a review, *Lab. Anim. Sci.*, 29, 645-651, 1979.

18. Bauck, L. and Bihun, C. Basic anatomy, physiology, husbandry, and clinical techniques, in *Ferrets, Rabbits, and Rodents: Clinical Medicine and Surgery*, Hillyer, E. V. and Quesenberry, K. E., eds., W. B. Saunders Company, Philadelphia, PA, 1997.

19. Dillon, W. G. and Glomski, C. A., The Mongolian gerbil: qualitative and quantitative aspects of the cellular blood picture, *Laboratory Animals*, 9, 283-287, 1975.

20. Harkness, J. and Wagner, J., in *The Biology and Medicine of Rabbits and Rodents*, Lea & Febiger, Philadelphia, PA, 1989.

21. Mays, A. Baseline hematological and blood biochemical parameters of the Mongolian gerbil, (*Meriones unguiculatus*), *Lab. Anim. Sci.*, 19, 838-842, 1969.

22. Melby, E. C. and Altman, N. H., *Handbook of Laboratory Animal Science, Volume II*. CRC Press, Inc., Cleveland, OH, 1974, pp. 365-436.

23. Mitruka, B. M. and Rawnsley, H. M., *Clinical, Biochemical, and Hematological Reference Values in Normal Experimental Animals*, Masson Publishing, New York, 1977.

24. Tomson, F. N. and Wardrop, K. J., Clinical chemistry and hematology, in *Laboratory Hamsters*, Van Hoosier, G. L. and McPherson, C. W., Academic Press, Inc. New York, 1987.

25. Wardrop, K. J. and Van Hoosier, G. L. The hamster, in *The Clinical Chemistry of Laboratory Animals*, Loeb, W. F. and Quimby, F. W., eds., Pergamon Press, New York, NY. 1988. pp. 31-39.

26. Reed, R. K., Jones, R. B., Bearg, D. W., Bedigian, and H. Paigen, B., Impact of room ventilation rates on mouse cage ventilation and microenvironment, *Contemporary Topics*, 36, 74-79, 1997.

27. National Research Council, *Guide For the Care and Use of Laboratory Animals*, Public Health Service, Washington, D.C., 1996.

28. Code of Federal Regulations (CFR), Title 9; Parts 1, 2, and 3 (Docket 89-30), *Federal Register*, vol. 54, No. 168, August 31, 1989, and CFR Part 3, (Docket No. 90-218), *Federal Register*, vol. 56, No. 32, February 15, 1991.

29. Harkness, J. and Wagner, J., in *The Biology and Medicine of Rabbits and Rodents*, Lea & Febiger, New York, 1995.

30. Arnold, C. E. and Estep, D. Q., Laboratory caging preferences in golden hamsters (*Mesocricetus auratus*), *Lab. Anim. Sci,*. 28: 232-38, 1994.

31. Balk, M. W. and Slater, G. M., Care and management, in *Laboratory Hamsters*, Van Hoosier, G. L. and McPherson, C. W., Academic Press, Orlando, FL, 1987, pp 61-68.

32. Ehle, F. R. and Warner, R. G., Nutritional implications of the hamster forestomach, *J. Nut.*, 108, 239-43, 1978.

33. Otken, C. C. and Scott, C. E., Feeding behavior in the Mongolian gerbil (*Meriones unguiculatus*), *Lab. Anim. Sci.*, 34, 181-184, 1984.

34. Battles, A. H., The biology, care, and diseases of the Syrian hamster, *Comp. Cont. Ed. Prac. Vet.*, 7, 825–828, 1985.

35. Steele, E., Odor recognition by male hamsters: discrimination of the hormonal state of females by odor from vaginal secretions, *J, Endo.*, 105, 255, 1985.

36. Animal Welfare Act, United States PL 89–544, 1996: P.L. 91–579, 1970; P.L. 94–279, 1976; PL 99–198, 1985 (The Food Security Act).

37. Health Research Extension Act, United States P.L. 99–158, 1985.

38. National Research Council, *Occupational Health and Safety in the Care and Use of Research Animals*, National Academy Press, Washington, D.C., 1997.

39. *Occupational Health and Safety in the Care and Use of Research Animals*, Committee on Occupational Safety and Health in Research Animal Facilities, ILAR Commission on Life Sciences, National Academy Press, Washington, D.C., 1997.

40. Fox J. G., Newcomer, C. E., and Rozmiarek, H., Selected zoonoses and other health hazards, in *Laboratory Animal Mecidine,* Fox, J. G., Cohen, B. J. and Loew, F. M., eds., Academic Press, Orlando, FL, 1984

41. Hugh-Jones, M. E., Hubbert, W. T., and Hagstad, H. V., *Zoonoses: Recognition, Control, and Prevention,* Iowa State University Press, Ames, IA, 1995.

42. Van Hoosier, G. L., and Ladiges, W. C., Biology and diseases of hamsters, in *Laboratory Animal Mecidine,* Fox, J. G., Cohen, B. J., and Loew, F. M., eds., Academic Press, Orlando, FL, 1984.

43. Drozdowicz, C. K., Bowman, T. A., Webb, M. L., and Lang, C. M., Effect of in-house transport on murine plasma corticosterone concentration and blood lymphocyte populations, *Amer. J. Vet. Res.*, 51, 1841-1846, 1990.

44. Landi, M. S., Kreider, J. W., Lanf, C. M., and Bullock, L. P. Effects of shipping on the immune function in mice, *Am. J. Vet. Res.*, 43, 1654-1657, 1982.

45. Cooper, D. M. and Gebhart, C. J., Comparative aspects of proliferative enteritis, *J. Amer. Vet. Med. Assoc.*, 212, 1446-1451, 1998.

46. Frisk, C. S., *Bacterial and Mycotic Diseases in Laboratory Hamsters*, Van Hoosier, G. L. and McPherson, C. W., Academic Press, Orlando, FL, 1987, pp. 112-128.

47. Parker, J. C., Ganaway, J. R., and Gillett, C. S., Viral diseases, in *Laboratory Hamsters*, Van Hoosier, G. L., and McPherson, C. W., Academic Press, Orlando, FL, 1987, pp. 95-106.

48. Wagner, J. E., *Parasitic Disease in Laboratory Hamsters*, Van Hoosier, G. L. and McPherson, C. W., Academic Press, Orlando, FL, 1987b, pp. 135-53.

49. Kellog, H. S. and Wagner, J. E., Experimental transmission of *Syphacia obvela*ta among mice, rats, hamsters and gerbils, *Lab. Anim. Sci.*, 32, 500-502, 1982.

50. Mezza, L. E., Quimby, F. W., Durham, S. K., Lewis, R. M., Characterization of spontaneous amyloidosis of Syrian hamsters using the potassium permanganate method, *Lab. Anim. Sci.*, 34, 376-380, 1984.

51. McMartin D. N., Spontaneous atrial thrombosis in aged syrian hamsters. I. Incidence and pathology, *Haemostasis*, 38, 447-456, 1977.

52. McMartin, D. N. and Dodds, U. J., Atrial thrombosis in the aged syrian hamster. Animal model of human disease, *Amer. J. Path.*, 107, 227-29, 1982.

53. Weschler, S.J. and Jones, J., Diagnostic exercise, *Lab. Anim. Sci.*, 34, 137-38, 1984.

54. Gleiser, C.A., A polycystic disease of hamsters in a closed colony, *Lab. Anim. Care*, 20, 923-929, 1970.

55. Bresnahan, J. F. Smith, G. D., and Lentsch, R. H., Nasal dermatitis in the Mongolian gerbil, *Lab. Anim. Sci.*, 33, 258-263, 1983.

56. Farrar, P. L., Opsomer, M. T., Kocen, J. A, and Wagner, J.E., Experimental nasal dermatitis in the Mongolian gerbil: effect of bilateral harderian gland adenectomy on development of facial lesions, *Lab. Anim. Sci.*, 38, 72-79, 1988.

57. Peckman, J. C., Cole, J. R., Chapman, W. L., Malone, J. B., McCall, J. W., and Thompson, P. E., Staphylcoccal derma-

titis in Mongolian gerbils (*Meriones unguiculatus*), *Lab. Anim. Sci.*, 24, 43-47, 1974.

58. Vincent, A. L., Rodrick, G. E., and Sodeman, W. A., The pathology of the Mongolian gerbil (*Meriones unguiculatus*): a review, *Lab. Anim. Sci.*, 29, 645-51, 1979.

59. Wightmann, S. R., Mann, P. C., and Wagner, J. E., Dihydrostreptomycin toxicity in the Mongolian gerbil, *Lab. Anim. Sci.*, 30, 71-75, 1980.

60. Smith, D. A. and Burgmannn, P., Formulary, in *Ferrets, Rabbits, and Rodents,* Hillyer, E. V. and Quesenberry, K. E., eds., Clinical Medicine and Surgery, W.B. Saunders Company, Philiadelphia, PA, 1997., pp 392-403.

61. Flecknell, P. A., *Laboratory Animal Anesthesia*, Academic Press, Ltd. London, U.K., 1987.

62. Hughes, H. C., Anesthesia of laboratory animals, *Lab. Anim.*, 15, 40-56, 1981.

63. Flecknell, P. A., and Mitchell, M. Injectable anesthetic techniques in two species of gerbil (*Meriones libycus* and *Meriones ungiculatus*), *Lab. Anim.*, 17, 118-122, 1983.

64. Lightfoot, W. E., II and Molinari, G.K., Comparison of ketamine and pentobarbital anestheisa in the Mongolian gerbil, *Am. J. Vet. Res.*, 39, 1061-1063, 198.

65. Clifford, D., Preanesthesia, anesthesia, analgesia and euthanasia, in *Laboratory Animal Mecidine*, Fox, J. G., Cohen, B.J., and Loew, F. M. (eds)., Academic Press, Orlando, FL, 1984.

66. Mason, D. E., Anesthesia, analgesia, and sedation for small mammals, in *Ferrets, Rabbits, and Rodents: Clinical Medicine and Surgery*, Hillyer, E. V. and Queensberry, K. E., W. B. Saunders Co, Philadelphia, PA, 378-397, 1997.

67. Knecht, C. D., Allen, A. R., Williams, D. J. Johnson, and J. H., *Fundamental Techniques in Veterinary Surgery*, W. B. Saunders Co., Philadelphia, PA, 1981.

68. Bojrab, M. J., Crane, S. W., and Arnoczky, S. P., *Current Techniques in Small Animal Surgery*, Lea & Febiger, Philadelphia, PA, 1983.

index

notes

notes

notes

notes

notes

T - #0582 - 101024 - C0 - 234/156/9 - PB - 9780849325663 - Gloss Lamination